Mike Dobbs & Linda Binek
4510 Cedar Point Ct.
Howell, MI 48843
P9-EEC-569

Electro-Optic and Acousto-Optic Scanning and Deflection

OPTICAL ENGINEERING

Series Editor

Brian J. Thompson
William F. May Professor of Engineering
and
Dean, College of Engineering and Applied Science
University of Rochester

1. Electron and Ion Microscopy and Microanalysis: Principles and Applications, *by Lawrence E. Murr*
2. Acousto-Optic Signal Processing: Theory and Implementation, *edited by Norman J. Berg and John N. Lee*
3. Electro-Optic and Acousto-Optic Scanning and Deflection, *by Milton Gottlieb, Clive L. M. Ireland, and John Martin Ley*

Other Volumes in Preparation

Electro-Optic and Acousto-Optic Scanning and Deflection

Milton Gottlieb
Westinghouse Research and Development Center
Pittsburgh, Pennsylvania

Clive L. M. Ireland
J. K. Lasers, Ltd.
Rugby, Warwickshire, England

John Martin Ley
Electro-Optic Developments, Ltd.
Basildon, Essex, England

MARCEL DEKKER, INC. New York and Basel

Library of Congress Cataloging in Publication Data

Gottlieb, Milton.
Electro-optic and acousto-optic scanning and deflection.

(Optical engineering ; v. 3)
Includes index.
1. Electrooptical devices. 2. Acoustooptical devices. 3. Optical scanners. 4. Light modulators. I. Ireland, Clive L. M., [date]. II. Ley, John Martin, [date]. III. Title. IV. Series: Optical engineering (Marcel Dekker, Inc.) ; v. 3.
TA1750.G67 1983 621.36'7 83-1867
ISBN 0-8247-1811-9

MARCEL DEKKER, INC.
270 Madison Avenue, New York, New York 10016

Current printing (last digit):
10 9 8 7 6 5 4 3 2 1

PRINTED IN THE UNITED STATES OF AMERICA

Preface

Any student of electro-optics or acousto-optics is aware of the wealth of published information available on the subject, but it is scattered in the numerous scientific journals of our time. This book attempts to cover sufficient electro- and acousto-optic theory for the reader to understand and appreciate the design and application of solid state optical deflectors. At the same time it is hoped that for the more experienced engineer it will serve as a useful reference book covering the most important work in this field of engineering.

Historically, the two techniques progressed independently with the investigation of the linear electro-optic (E-O) effect by Pöckels in 1894 and the acousto-optic (A-O) effect by Debye and Sears in 1932. Interestingly, the development of Sonar during World War II involved a range of piezoelectric crystals, including potassium dihydrogen phosphate (KDP) and ammonium dihydrogen phosphate (ADP), which were also used to produce the first E-O modulators. This interrelation continued with lithium niobate which has been used in many E-O deflectors and also used for the transducers in most A-O deflectors. These associations relate more to fundamental material properties than to the electro- and acousto-optic effects and device techniques with which this book is primarily concerned. It is on the latter basis that the two subjects have been treated independently and presented in separate sections. Something is lost and something gained; it is hoped that for most readers the choice will be considered the right one.

The development of the laser in 1960 stimulated the demand for methods to switch, modulate, and deflect light for the great variety of applications that were foreseen. The principal mechanisms then proposed for producing fast, high resolution deflection of laser light involved electro-optic and acousto-optic interactions, and it is to a large extent these mechanisms and their application to deflection that are discussed in this book.

Optimism was supported at that time by the success obtained with the E-O material potassium tantalate-niobate (KTN), first grown in 1964. A paper published by Chen et al. in 1966 concluded, "light modulators with several hundred MHz bandwidth should provide 100% modulation with dissipated powers of several milliwatts and reactive powers of a few watts. In addition, an analog beam deflector (double compensated prism) capable of producing several hundred resolvable spots appears to be possible." Unfortunately, progress with KTN was curtailed since its other crystal properties were found to be unsatisfactory and because of severe crystal growth problems. Nevertheless, E-O deflectors and scanners have been fabricated from other less sensitive, but more tractable materials and successfully employed in a wide range of applications.

The phenomena underlying the interaction of light waves with sound waves were largely understood by the mid-1930s but remained as scientific curiosities, having no practical significance until the 1960s. During this period, new technologies led to the production of high efficiency, wideband transducers capable of operating to several GHz, and at the same time small, high power, wideband solid state amplifiers became available to drive these transducers. Finally, a number of new synthetic acousto-optic crystals with low drive power requirements and low acoustic losses at high frequencies were produced, enabling the realization of practical A-O devices both for the modulation and deflection of light.

By the mid-1970s acousto-optic modulators had achieved a similar user position to the electro-optic modulator and, in default of research workers producing a more efficient electro-optic material, acousto-optic deflection provided the only technique for high resolution solid state optical scanning. The acousto-optic deflector has now been developed to the point where it is very often the only possible device to satisfy many practical requirements. In particular, in the world of information storage and presentation, the bulk acousto-optic deflector and its counterpart light source, the gas laser, currently provide the only available random access high speed writing facility. It will be very interesting to see if the performance of the A-O deflector will be matched or surpassed in the coming years by E-O devices based on new E-O materials. The technology for each type of device is still in its infancy and the pressure to develop faster and more flexible deflection systems will undoubtedly push it forward rapidly in the next ten years or so. It is the authors' sincere hope that this review of the work carried out to date will be found a useful guide to those involved in this future advance.

Milton Gottlieb
Clive L. M. Ireland
John Martin Ley

Contents

Series Introduction

Optical science, engineering, and technology have grown rapidly in the last decade so that today optical engineering has emerged as an important discipline in its own right. This series is devoted to discussing topics in optical engineering at a level that will be useful to those working in the field or attempting to design systems that are based on optical techniques or that have significant optical subsystems. The philosophy is not to provide detailed monographs on narrow subject areas but to deal with the material at a level that makes it immediately useful to the practicing scientist and engineer. These are not research monographs, although we expect that workers in optical research will find them extremely valuable.

Volumes in this series cover those topics that have been a part of the rapid expansion of optical engineering. The developments that have led to this expansion include the laser and its many commercial and industrial applications, the new optical materials, gradient index optics, electro- and acousto-optics, fiber optics and communications, optical computing and pattern recognition, optical data reading, recording and storage, biomedical instrumentation, industrial robotics, integrated optics, infrared and ultraviolet systems, etc. Since the optical industry is currently one of the major growth industries this list will surely become even more extensive.

Brian J. Thompson
University of Rochester
Rochester, New York

part I

ELECTRO-OPTIC DEFLECTORS

CLIVE L. M. IRELAND / J. K. Lasers Ltd., Rugby, Warwickshire, England
JOHN MARTIN LEY / Electro-Optic Developments, Ltd., Basildon, Essex, England

The main stimulus for the development of electro-optic (E-O) devices has been the invention of the laser. Although the observation of the electric field dependence of the optical properties of certain materials was reported by Kerr and Pockels around the turn of the century, it was the development of the laser, with its high-intensity and near diffraction-limited beam, that allowed practicable application of these E-O effects in optical switching, modulating, and deflecting devices.

By 1960, when the first laser was operated, a considerable body of information had been collected on the E-O properties of many materials. In particular, Zwicker and Scherrer (1943) had reported on the materials potassium dihydrogen phosphate (KH_2PO_4) and deuterated potassium dihydrogen phosphate (KD_2PO_4), which are still among the most important for use in commercial E-O devices. As a result of the impetus received from the development of the laser, a wealth of information on the E-O properties of many other crystals and liquids has since been obtained. Unfortunately, as yet, few materials exhibit a strong enough E-O effect to make them of use in practicable E-O devices--the E-O industry still awaits the "ideal" material. Nevertheless, E-O deflectors have been fabricated and successfully employed in a wide range of applications, albeit often with somewhat complicated geometry or less than ideal performance.

Recently, a detailed theoretical understanding of the origin of the E-O effect in terms of the crystal atomic structure and electronic bonding has been developed and applied to the known simpler E-O crystal structures with reasonable success. Although it is not currently possible to predict with certainty the magnitude of the E-O effect in a particular crystal, this work gives cause for optimism that, before long, bulk E-O crystal engineering will be possible. Undoubtedly, in the future, new crystals showing a large E-O effect will be identified and grown to usable size. The development of a more efficient electro-optic material would greatly accelerate the application of bulk electro-optic deflectors.

In the following chapters, the basic E-O effect is reviewed, the properties and selection of E-O materials discussed, the principles of operation of particularly important deflectors examined, and the performance of specific types compared. Finally, a brief account of likely future developments and applications is given.

1

Theory of the Electro-Optic Effect

1.1 THE ELECTRO-OPTIC EFFECT

An understanding of the nature of the electro-optic (E-O) effect in materials is a prerequisite to understanding the principles of operation of E-O deflectors. In all materials an applied electric field results in a change in polarization. These changes are generally very small and except in a relatively few cases do not result in a measurable change in the optical properties of the material. Even if a measurable change results, the smallness of the effect restricts the number of materials from which practical E-O switching, modulating, or deflecting devices can be made.

Both the applied electric field $\underline{E}$ and the induced polarization $\underline{P}$ are vector quantities and, in general, are not in the same direction. Consequently, they are related by a tensor equation of the following general form (see, for example, Yariv [1]):

$$\underline{P} = \varepsilon_0(\psi^{(1)} \cdot \underline{E} + \psi^{(2)}:\underline{E}^2 + \psi^{(3)}:\underline{E}^3 + \cdots) \tag{1.1}$$

Here ε_0 is the free-space permittivity. The coefficient $\psi^{(1)}$ in (1.1) is the susceptibility used to describe the familiar linear optic phenomena of reflection and refraction. Although the higher (nonlinear) term with coefficients $\psi^{(2)}$, $\psi^{(3)}$, etc., can be used to describe the behavior of the media for low-frequency fields, that is, those below the frequency of the optical lattice modes of the material [2], use of the equation cast in this form is generally restricted to cases where all the field components are at high (optical) frequency. For a description of the E-O properties of the material at low frequencies it is usual to express directly the change in the refractive index n as a function of $\underline{E}$ by

$$\frac{1}{n^2} = \frac{1}{n_0^2} + r \cdot \underline{E} + h:\underline{E}^2 + \cdots \tag{1.2}$$

Here n_0 is the zero field value of the refractive index and r and h are linear and quadratic E-O coefficients, respectively.

The reason for describing the field-induced changes in the optic properties of the material through the parameter $1/n^2$ is related to the fact that the most useful E-O materials are crystalline. In the following section it will be shown how an equation with coefficients of $1/n^2$ can be used to characterize electromagnetic (EM) wave propagation in a general anisotropic crystalline material. With (1.2), use of the equation is extended to cover the optical properties of the material when it is subject to an external applied field.

1.2 EM WAVE PROPAGATION IN A CRYSTAL

For the general case of an anisotropic crystal the electric displacement $\underline{D}$, resulting from an applied field $\underline{E}$, is expressed by the second rank tensor $\varepsilon_{k\ell}$ through the relationship

$$D_k = \varepsilon_{k\ell}\underline{E}_\ell \tag{1.3}$$

Here the subscripts refer to a Cartesian coordinate $(k, \ell = x_1, x_2, x_3)$, where x_1, x_2, x_3 are fixed with respect to the crystal axes and the convention of summation over repeated indices is observed. By equating the net power flow into the material (given by Poynting's vector) with the rate of change of electrical stored energy, it can be shown [1,3] that $\varepsilon_{k\ell} = \varepsilon_{\ell k}$. Consequently, $\varepsilon_{\ell k}$ has, in general, only six independent elements.

Noting that the stored electrical energy density w in the crystal is given by

$$w = \frac{1}{8\pi}\,\underline{E}\cdot\underline{D} = \frac{1}{8\pi}E_k\varepsilon_{k\ell}E_\ell \tag{1.4}$$

we therefore have

$$8\pi w = \varepsilon_{x_1x_1}E_{x_1}^2 + \varepsilon_{x_2x_2}E_{x_2}^2 + \varepsilon_{x_3x_3}E_{x_3}^2 + 2\varepsilon_{x_2x_3}E_{x_2x_3} + 2\varepsilon_{x_1x_3}E_{x_1}E_{x_2} + 2\varepsilon_{x_1x_2}E_{x_1x_2} \tag{1.5}$$

With a suitable choice of new coordinate axes, the last three terms in (1.5) can be eliminated. The new axes are called the *principal dielectric axes* of the crystal, and the change in axes is a principal axes transformation. In the new coordinate system (x,y,z), Eq. (1.5) becomes

$$8\pi w = \varepsilon_xE_x^2 + \varepsilon_yE_y^2 + \varepsilon_zE_z^2 \tag{1.6}$$

In this coordinate system the tensor $\varepsilon_{k\ell}$ is diagonal, so that $D_x = \varepsilon_x E_x$, etc. Thus (1.6) can be written as

$$8\pi w = \frac{D_x^2}{\varepsilon_x} + \frac{D_y^2}{\varepsilon_y} + \frac{D_z^2}{\varepsilon_z} \tag{1.7}$$

Consequently, the constant energy density surfaces (w = constant) in the space D_x, D_y, D_z are ellipsoids. Here ε_x, ε_y, and ε_z are the principal dielectric constants of the crystal. Further, if we replace $D_k/\sqrt{8\pi w}$ by x_k and define the principal indices of refraction n_x, n_y, and n_z of the crystal by $n_k^2 = \varepsilon_k$ (k = x, y, z), Eq. (1.7) becomes

$$\frac{x^2}{n_x^2} + \frac{y^2}{n_y^2} + \frac{z^2}{n_z^2} = 1 \tag{1.8}$$

This is the equation of a general ellipsoid with axes parallel to the x, y, and z directions whose respective lengths are $2n_x$, $2n_y$, and $2n_z$. The ellipsoid, which is known as the index ellipsoid, indicatrix, or sometimes the ellipsoid of wave normals, characterizes EM wave propagation in the crystal. If all three principal refractive indices are equal, the crystal is isotropic; if only two are equal, it is termed uniaxial; and if none are equal, it is called biaxial.

The major use of the index ellipsoid is to determine the value of the refractive index associated with the polarization (direction of $\underline{D}$) of the two normal modes of a plane wave propagating along an arbitrary direction $\underline{s}$ in the crystal. A plane normal to the direction $\underline{s}$ and passing through the center of the ellipsoid cuts it in an ellipse whose major and minor axes define the directions of the two independent planes of polarization. The half-lengths of these axes are equal in value to the two indices of refraction.

As an example of the use of the index ellipsoid we shall consider beam propagation in a uniaxial crystal. In this case (1.8) becomes

$$\frac{x^2}{n_o^2} + \frac{y^2}{n_o^2} + \frac{z^2}{n_e^2} = 1 \tag{1.9}$$

where n_o and n_e designate the two principal (ordinary and extraordinary) refractive indices, respectively, and where the convention of taking the z crystal axis as the axis of symmetry has been adopted. For the particular case of a beam propagating at an angle θ of $\pi/2$ to the z crystal axis, the plane normal to $\underline{s}$ through the center of the ellipsoid intersects the ellipsoid to form an ellipse with semiaxes of length n_o and n_e. This is illustrated in Fig. 1.1a. The more general

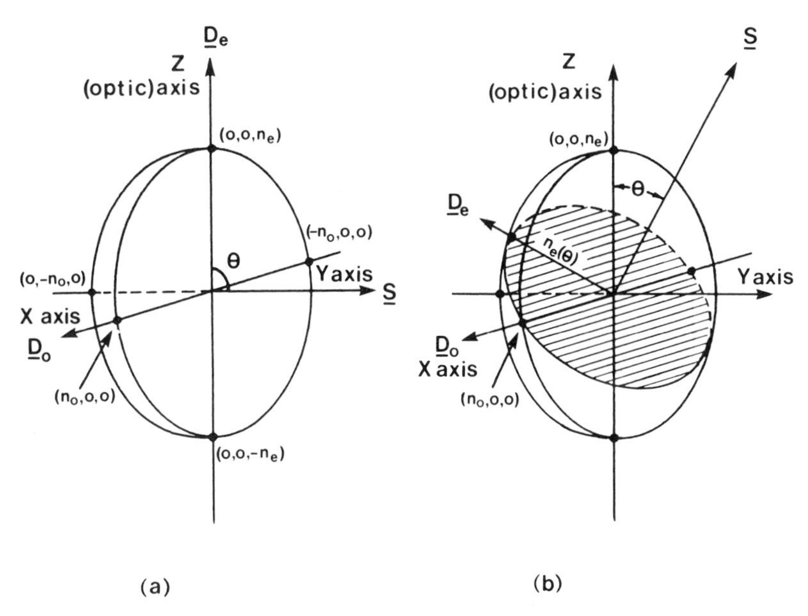

Figure 1.1 Use of the index ellipsoid for finding the refractive index for the two normal mode planes of polarization associated with an EM wave propagating in a uniaxial crystal. (a) In this simple example, the direction of beam propagation $\underline{s}$ is at $\theta = \pi/2$ to the Z (optic) axis, and the two refractive indices are n_o and n_e. (b) As in (a) but the more general case where $\theta \neq \pi/2$. Here the refractive indices (the semi-axes of the hatched elliptic cross-section) are n_o and $n_e(\theta)$, where $n_e(\theta)$ is given by (1.10).

case where $\theta \neq \pi/2$ is shown in Fig. 1.1b. Here the length of the semi-axes of the ellipsoid are n_o and $n_e(\theta)$, where $n_e(\theta)$ is given by

$$\frac{1}{n_e^2(\theta)} = \frac{\cos^2\theta}{n_o^2} + \frac{\sin^2\theta}{n_e^2} \tag{1.10}$$

Consequently, the extraordinary refractive index varies from $n_e(\theta) = n_o$ for $\theta = 0°$ to $n_e(\theta) = n_e$ for $= \pi/2$. Crystals with $n_e > n_o$ are called positive uniaxial and with $n_e < n_o$ are called negative uniaxial. Useful examples of these two types of crystal are quartz and calcite, respectively. It is clear that any of these birefringent materials, cut in the form of a prism and orientated in a manner similar to that in Fig. 1.1a, would act as a good polarizer, since it would introduce an angular separation between the two independent polarizations of the beam. Since calcite is available in large, high-optic-quality pieces and has a large birefringence, it is very commonly used as a polarizer material. In particular, it has found wide use in this role in digital-type deflectors.

The majority of crystals occurring in nature are biaxial. Positive crystals include sulfur (S), topaz $Al_2SiO_4(F,OH)$, and turquoise $CuO\,3Al_2O_3\,2P_2O_5\;9H_2O$, and negative crystals include mica $KH_2Al_3(SO_4)_3$, argonite $CaO(CO)_2$, and stibnite Sb_2S_3. For the purpose of determining the refractive indices in these materials, the second (azimuthal) direction ϕ of the beam propagation, relative to the crystal axes, has to be defined since the index ellipsoid is no longer symmetric about z. These crystals belong to the orthorhombic, monoclinic, or triclinic systems. Their dielectric axes may or may not be determined by crystal symmetry and consequently can be wavelength dependent.

1.3 THE LINEAR ELECTRO-OPTIC EFFECT

The preceding discussion has shown how beam propagation in an anisotropic crystal can be characterized by the index ellipsoid. In a general (i.e., nonprincipal) coordinate system the equation of this ellipsoid can be obtained from (1.5), viz.,

$$\frac{1}{n_1^2}x_1^2 + \frac{1}{n_2^2}x_2^2 + \frac{1}{n_3^2}x_3^2 + \frac{1}{n_4^2}x_2x_3 + \frac{1}{n_5^2}x_1x_3 + \frac{1}{n_6^2}x_1x_2 = 1 \qquad (1.11)$$

Interchanging the indices in (1.5) leaves the equation unchanged; consequently, the contraction has been made here of writing $1 \leftrightarrow x_1x_1$, $2 \leftrightarrow x_2x_2$, $3 \leftrightarrow x_3x_3$, $4 \leftrightarrow x_2x_3$, $5 \leftrightarrow x_1x_3$, and $6 \leftrightarrow x_1x_2$. As (1.11) is a general equation characterizing beam propagation in a crystal, it is natural that changes in the optic properties of the crystal due to an applied field should be expressed as a change in its coefficients. The linear E-O effect (Pockels effect) relates the linear variation of these coefficients to the field and is described by the E-O tensor r_{ij} through the relationship

$$\Delta\left(\frac{1}{n^2}\right)_i = r_{ij}E_j \qquad (1.12)$$

where $\Delta(1/n^2)_i$ is the change in the coefficient $(1/n^2)_i$ and where, as usual, we sum over repeated indices.

In centrosymmetric crystals, that is, those possessing an inversion symmetry, the optical characteristics must remain the same when the sign of the applied field is reversed. It follows from (1.12) that $\Delta(1/n^2)_i = r_{ij}(-E_j)$, which can be satisfied only if $r_{ij} = 0$. Consequently, the linear E-O effect is found only in crystals lacking an inversion symmetry. The lack of inversion symmetry is also the prerequisite for the existence of a linear piezoelectric effect which is described by a 3 × 6 tensor d_{ji}. The form of both the r_{ij} and d_{ji} tensors is determined uniquely by the point group symmetry of the crystal. The form of the r_{ij} tensor is obtained from the d_{ji} tensor by conjugation, that is, by replacing terms d_{ji} by r_{ij}. This converts the 3 × 6 piezoelectric tensor into the 6 × 3 form describing the linear E-O effect. Although a large piezoelectric effect does not necessarily imply a large E-O effect, the relationship has been used widely to aid the selection of potentially useful E-O materials [4].

By combining (1.11) and (1.12), the equation of the index ellipsoid in the presence of the field E_j is obtained; that is,

$$\left(\frac{1}{n_1^2} + r_{1j}E_j\right)x_1^2 + \left(\frac{1}{n_2^2} + r_{2j}E_j\right)x_2^2 + \left(\frac{1}{n_3^2} + r_{3j}E_j\right)x_3^2 + 2\left(\frac{1}{n_4^2} + r_{4j}E_J\right)x_2x_3$$

$$+ 2\left(\frac{1}{n_5^2} + r_{5j}E_j\right)x_1x_3 + 2\left(\frac{1}{n_6^2} + r_{6j}E_j\right)x_1x_2 = 1 \tag{1.13}$$

In general, it is not necessary to completely determine the surface represented by this equation. To find the two independent refractive indices for the beam, it is only necessary to determine the lengths of the semiaxes of the ellipse, normal to the direction of beam propagation $\underline{s}$, through the center of the ellipsoid. For example, if $\underline{s}$ and the applied field $\underline{E}_j$ are directed along the x_2 and x_3 axes, respectively, (1.13) reduces to

$$\left(\frac{1}{n_1^2} + r_{13}E_3\right)x_1^2 + \left(\frac{1}{n_3^2} + r_{33}E_3\right)x_3^2 + 2\left(\frac{1}{n_5^2} + r_{53}E_3\right)x_1x_3 = 1 \tag{1.14}$$

In the special case where the crystal is orientated so that x_1, x_2, and x_3 correspond to the crystal axes x, y, and z, respectively, then we also have $1/n_4^2 = 1/n_5^2 = 1/n_6^2 = 0$. This is a geometry that has been used widely in deflector designs. Consequently,

$$\left(\frac{1}{n_x^2} + r_{13}E_z\right)x^2 + \left(\frac{1}{n_z^2} + r_{33}E_z\right)z^2 + 2r_{53}E_z xz = 1 \tag{1.15}$$

Depending on the symmetry of the particular crystal used, n_x and n_z may or may not be equal, and some r_{ij} terms may be equal or vanish.

Ignoring the term in r_{53} for the moment and making use of the fact that $\Delta(1/n^2) = -2\Delta n/n^3$ for small Δn, we see that (1.15) represents

Table 1.1 Properties of Ten Important E-O Materials Near Room Temperature[a]

Material	Transition temp. (K)	Principle r_{ij}'s ($\times 10^{-12}$ m/V)		Refractive indices ($\sim$ 600 nm)	Dielectric const. ($\varepsilon/\varepsilon_0$)	Transmission range (μm)
KDP (KH_2PO_4)	123	$r^T_{41} \sim 8.6$	$r^T_{63} \sim 10.4$	$n_o \sim 1.51$	$\varepsilon^T \parallel C \sim 20$	$\sim$ 0.2-1.5
				$n_e \sim 1.47$	$\varepsilon^S \parallel C \sim 20$	
		$r^S_{41} \sim$	$r^S_{63} \sim 9.0$		$\varepsilon^T \parallel C \sim 42$	
					$\varepsilon^S \parallel C \sim 44$	
KD*P (KD_2PO_4)	222	$r^T_{41} \sim 9.0$	$r^T_{63} \sim 25$	$n_o \sim 1.51$	$\varepsilon^T \parallel C \sim 50$	$\sim$ 0.2-2.1
					$\varepsilon^S \parallel C \sim 48$	
		$r^S_{41} \sim$	$r^S_{63} \sim 23$	$n_e \sim 1.47$	$\varepsilon^T \perp C \sim$	
					$\varepsilon^S \perp C \sim 57$	
ADP ($NH_4H_2PO_4$)	148	$r^T_{41} \sim 24$	$r^T_{63} \sim 8.5$	$n_o \sim 1.52$	$\varepsilon^T \parallel C \sim 15$	$\sim$ 0.2-1.4
					$\varepsilon^S \parallel C \sim 14$	
		$r^S_{41} \sim$	$r^S_{63} \sim 4.5$	$n_e \sim 1.48$	$\varepsilon^T \perp C \sim 75$	
					$\varepsilon^S \perp C \sim 74$	

KDA (KH_2AsO_4)	97	$r^T_{41} \sim 12.5$	$r^T_{63} \sim 11$	$n_o \sim 1.57$	$\varepsilon^T \parallel C \sim 21$	$\sim 0.3\text{-}1.6$
					$\varepsilon^S \parallel C \sim 19$	
		$r^S_{41} \sim$	$r^S_{63} \sim$	$n_e \sim 1.52$	$\varepsilon^T \perp C \sim 54$	
					$\varepsilon^S \perp C \sim 53$	
RDP (RbH_2PO_4)	147	$r^T_{41} \sim 9.1$	$r^T_{63} \sim 15.5$	$n_o \sim 1.50$	$\varepsilon^T \parallel C \sim 25$	$\sim 0.2\text{-}1.4$
					$\varepsilon^S \parallel C \sim$	
		$r^S_{41} \sim$	$r^S_{63} \sim 14$	$n_e \sim 1.47$	$\varepsilon^T \perp C \sim 40$	
					$\varepsilon^S \perp C \sim$	
$LiNbO_3$	1470	$r^T_{33} \sim 32$	$r^T_{42} \sim 32$	$n_o \sim 2.30$	$\varepsilon^T \parallel C \sim 32$	$\sim 0.4\text{-}5.0$
					$\varepsilon^S \parallel C \sim 28$	
		$r^S_{33} \sim 31$	$r^S_{42} \sim 28$	$n_e \sim 2.21$	$\varepsilon^T \perp C \sim 80$	
					$\varepsilon^S \perp C \sim 44$	
$LiTaO_3$	890	$r^T_{33} \sim 22$	$r^T_{42} \sim$	$n_o \sim 2.18$	$\varepsilon^T \parallel C \sim 45$	$\sim 0.4\text{-}6.0$
					$\varepsilon^S \parallel C \sim 43$	
		$r^S_{33} \sim 30$	$r^S_{42} \sim 20$	$n_e \sim 2.19$	$\varepsilon^T \perp C \sim 51$	
					$\varepsilon^S \perp C \sim 41$	

Table 1.1 (Continued)

Material	Transition temp. (K)	Principle r_{ij}'s ($\times 10^{-12}$ m/V)		Refractive indices ($\sim$ 600 nm)	Dielectric const. ($\varepsilon/\varepsilon_0$)	Transmission range (μm)
$BaTiO_3$	390	$r_{33}^T \sim 108$	$r_{42}^T \sim 1640$	$n_o \sim 2.44$	$\varepsilon^T \parallel C \sim 120$	
					$\varepsilon^S \parallel C \sim 60$	
		$r_{33}^S \sim 23$	$r_{42}^S \sim 820$	$n_e \sim 2.37$	$\varepsilon^T \perp C \sim 4000$	
					$\varepsilon^S \perp C \sim 2300$	
CuCl		$r_{41}^T \sim 3.6$	$r_{41}^S \sim 2.4$	$n_o \sim 2.0$	$\varepsilon^T \sim 10.0$	$\sim$ 0.4-19.0
					$\varepsilon^S \sim 7.5$	
ZnTe		$r_{41}^T \sim 4.4$	$r_{41}^S \sim 4.3$	$n_o \sim 2.9$	$\varepsilon^T \sim 10$	
					$\varepsilon^S \sim 0$	

[a] Data are from references cited in the text. Agreement between authors for E-O coefficients is typically $\sim \pm 15\%$.

an ellipse with semiaxes along x and z equal to $n_x + \Delta n_x$ and $n_z + \Delta n_z$, respectively, where

$$\Delta n_x = \frac{-n_x^3}{2} r_{13} E_z \quad \text{and} \quad \Delta n_z = \frac{-n_z^3}{2} r_{33} E_z \tag{1.16}$$

These are the increments in refractive index, resulting from the application of the field E_z, for the x and z polarization components of the beam, respectively.

The term in xz in (1.6) has the effect of rotating the axes of the elliptical cross-section by an angle α, where

$$\tan 2\alpha = \frac{2r_{53}E_z}{1/n_1^2 - 1/n_2^2 + (r_{13} - r_{33})E_z} \tag{1.17}$$

In crystals where the natural birefringence is appreciable, that is, $|1/n_1^2 - 1/n_2^2| \gg (r_{13} - r_{33})E_z$, α is very small, and the principal axes lie close to the crystal axes. The lengths of the principal axes (and hence the refractive indices) are changed only by an amount of second order in $R_{53}E_z$ that is usually negligible. However, the radius vector of the ellipse at $\pi/4$ to x and z is changed linearly in E_z by an amount $n'^3/2r_{53}E_z$, where n' is an effective index intermediate between the principal indices [4,5].

If the particular crystal used in the example was, say, the important ferroelectric material $LiNbO_3$, then $r_{53} = 0$, and the principal axes would remain the crystal axes. For this material $n_x \neq n_z$, and so, cut in the form of a prism, it behaves as a polarizer like the calcite example considered earlier. But unlike the earlier example, the results expressed by (1.16) show that, through the linear E-O effect, the power of the prism can be changed linearly with applied field. This is the basic principle behind the operation of analog deflectors that use the linear E-O effect.

There is an additional effect that requires consideration when designing E-O devices. In E-O crystals, the application of an electric field leads, via piezoelectric and/or electrostrictive coupling (effects proportional to E and E^2, respectively), to strain which can distort the index ellipsoid. In the particular case of strain resulting from the inverse piezoelectric effect (usually the dominant effect for linear E-O crystals), it can be shown that these refractive index changes depend on crystal symmetry in the same way as the E-O effect itself. Hence,

$$\Delta\left(\frac{1}{n^2}\right)_i = p_{ij}^e S_j + r_{ik}^s E_k \tag{1.18}$$

Here k runs from $1 \rightarrow 3$, but i and j are the reduced indices discussed earlier and run from $1 \rightarrow 6$. S_j is the strain tensor, and p_{ij}^e is the strain-optic (or elasto-optic) tensor component measured under the conditions of constant electric field. r_{ik}^S is the E-O coefficient measured at constant strain. The superscripts are used to indicate these two conditions. If the applied electric field is at high frequency, inertia prevents the material straining macroscopically. The crystal is therefore *clamped* and operates under the condition of constant strain. In this case, the first term on the right-hand side of (1.18) is zero. For low-frequency modulation the crystal is *free*, and the effect of the strain-optic tensor components, in general, cannot be ignored. The terms representing this effect are usually incorporated into the E-O coefficients, which at low frequency are generally written r_{ik}^T to indicate that they are applicable to conditions of constant stress. For the case where the strain is wholly piezoelectric in origin, the high- and low-frequency E-O coefficients are related by

$$r_{ik}^T = r_{ik}^S + p_{ij}d_{kj} \tag{1.19}$$

in which p_{ij} and d_{kj} are the strain-optic and piezoelectric tensor coefficients, respectively.

The E-O coefficients of a large number of materials, under conditions of constant strain and constant stress, have been determined by many workers. Useful reviews of published data on E-O materials include Kaminow and Turner [4], Pressley [6], Milek and Neuberger [7], and Kuz'minour et al. [8]. Table 1.1 lists some of the more important materials and their main optic and E-O properties.

1.4 THE QUADRATIC ELECTRO-OPTIC EFFECT IN CRYSTALS

In centrosymmetric crystals the linear E-O effect vanishes, but both noncentrosymmetric and centrosymmetric crystals can exhibit a quadratic effect. If the quadratic term dominates, Eq. (1.2) reduces to

$$\Delta\left(\frac{1}{n^2}\right)_{ij} = h_{ijk\ell}E_kE_\ell \tag{1.20}$$

where h is a fourth-rank tensor and i, j, k, and ℓ are Cartesian coordinates. Again we sum over repeated indices. Usually this equation is recast with the refractive index expressed in terms of the induced polarization:

$$\Delta\left(\frac{1}{n^2}\right)_{ij} = g_{ijk\ell}P_kP_\ell \tag{1.21}$$

where $P_\ell = \psi_{\ell g}E_g$. Consequently,

$$h_{ijsg} = g_{ijk\ell}\psi_{ks}\psi_{\ell g} \tag{1.22}$$

The refractive index change expressed as a function of the polarization is generally preferred for expressing the quadratic E-O effect, since experiment has shown that the tensor components in (1.21) are nearly independent of temperature. In contrast, those in (1.20) embody the coefficients of the susceptibility tensor, which can be strongly temperature dependent, particularly for E-O materials that are operated near a phase transition.

Since in (1.21) both ij and $k\ell$ are interchangeable, the tensor elements $g_{ijk\ell}$ can be written in contracted notation as $g_{\ell m}$, where ℓ represents ij and m represents $k\ell$, and both ℓ and m run from $1 \rightarrow 6$, for example $g_{x_1x_2x_3x_1} \leftrightarrow g_{65}$. g is, in general, a 6×6 tensor identical in form to that of the strain-optic tensor discussed earlier. As with the linear E-O effect, two or more coefficients $g_{\ell m}$ may be equal, or terms vanish, depending on the crystal symmetry. For example, in the important cubic crystals with point group symmetry m3m, g is given by [6]

$$g = \begin{bmatrix} g_{11} & g_{12} & g_{12} & 0 & 0 & 0 \\ g_{12} & g_{11} & g_{12} & 0 & 0 & 0 \\ g_{12} & g_{12} & g_{11} & 0 & 0 & 0 \\ 0 & 0 & 0 & g_{44} & 0 & 0 \\ 0 & 0 & 0 & 0 & g_{44} & 0 \\ 0 & 0 & 0 & 0 & 0 & g_{44} \end{bmatrix} \tag{1.23}$$

Consequently, the refractive index changes expressed by (1.21) are

$$\Delta\left(\frac{1}{n^2}\right)_1 = \begin{bmatrix} g_{11} & g_{12} & g_{12} & 0 & 0 & 0 \\ g_{12} & g_{11} & g_{12} & 0 & 0 & 0 \\ g_{12} & g_{12} & g_{11} & 0 & 0 & 0 \\ 0 & 0 & 0 & g_{44} & 0 & 0 \\ 0 & 0 & 0 & 0 & g_{44} & 0 \\ 0 & 0 & 0 & 0 & 0 & g_{44} \end{bmatrix} \begin{bmatrix} P_{x_1}^2 \\ P_{x_2}^2 \\ P_{x_3}^2 \\ P_{x_2}P_{x_3} \\ P_{x_1}P_{x_3} \\ P_{x_1}P_{x_2} \end{bmatrix} \tag{1.24}$$

In the absence of an applied field, cubic crystals are isotropic, and the index ellipsoid is a sphere of radius n_0. If we take for this example a field along the z crystal axis (i.e., $P_x = P_y = 0$), then the index ellipsoid becomes

$$\left(\frac{1}{n_0^2} + g_{12}P_z^2\right)(x^2 + y^2) + \left(\frac{1}{n_0^2} + g_{11}P_z^2\right)z^2 = 1 \tag{1.25}$$

In this case, the principal axes remain x, y, z, that is, aligned with the crystal axes. The semiaxes of the ellipsoid that give the new refractive indices become

$$n_x = n_y = n_0 - \frac{1}{2}n_0^3 g_{12} P_z^2 \quad \text{and} \quad n_z = n_0 - \frac{1}{2}n_0^3 g_{11} P_z^2 \tag{1.26}$$

For the particular case of a beam propagating in the x direction and polarized in the z direction, the refractive index change due to the quadratic E-O effect would therefore be

$$\Delta n_z = \frac{1}{2}n_0^3 g_{11}\left(\frac{\varepsilon_z - 1}{4\pi}\right)^2 E_z^2 \tag{1.27}$$

where P_z has been written as $\psi E_z = |(\varepsilon_z - 1)/4\pi|E_z$.

Some of the largest changes in refractive index occur in crystals of this m3m symmetry group operated in the cubic phase above their transition temperatures. Although there is little difference in the g_{11} coefficients between these crystals, the dependence of ε on the nearness of the operating (room) temperature to the transition temperature causes some to exhibit a much larger refractive index change than others. These crystals are obvious candidates for possible use in analog deflectors.

1.5 THE QUADRATIC ELECTRO-OPTIC EFFECT IN LIQUIDS

Refractive index changes proportional to the square of the applied field are permitted by symmetry in all materials. Besides the crystals discussed in the last section, liquids that are strongly polar are of particular E-O interest since they can exhibit a high anisotropic, optic polarizability. By applying a strong external field, the molecules of these substances partially align with the field, causing the bulk material to become birefringent. Due to the thermal motion of the molecules, the alignment, and hence the birefringence, is temperature dependent.

The direction of main polarizability is usually nearly parallel to the dipole moment in these molecules. Consequently, the component of a beam polarized parallel to the main polarizability of the molecule (i.e., the applied field direction) sees an increased refractive index relative to that of the orthogonal polarization. This effect, which was observed by Kerr in glass and other materials, is generally described by the following simple equation:

$$n_p - n_s = B\lambda E^2 \tag{1.28}$$

Here λ is the vacuum wavelength of the beam, B is the Kerr constant for the material, and n_p and n_s are the parallel and orthogonal refractive index components, respectively.

A variety of Kerr substances have been investigated by several workers. See, for example, Lee and Hauser [9] and Kruger et al. [10]. Most of these substances are liquids, but some are solids. One of the earliest polar liquids investigated was nitrobenzene, and it is still the most popular for E-O applications. Switches using this liquid have been incorporated in digital beam deflectors.

REFERENCES

1. A. Yariv, *Quantum Electronics*, Wiley, New York (1967).
2. G. D. Boyd and D. A. Kleinman, *J. Appl. Phys. 39*:3597 (1968).
3. M. Born and E. Wolf, *Principles of Optics*, (5th ed.), Pergamon, London (1975).
4. I. P. Kaminow and E. H. Turner, *Proc. IEEE 54*:1374 (1966).
5. M. J. Dore, *SRDE Report 68022*, HMSO (Oct. 1968), p. 31.
6. R. J. Pressley, (ed.), *Handbook of Lasers*, The Chemical Rubber Co., Cleveland (1971).
7. J. T. Milek and M. Neuberger, Linear E-O modulator materials, in *Handbook of Electronic Materials*, Vol. 8, IFI/Plenum, New York (1972).
8. Yu S. Kuz'minour, V. V. Osiko, and A. M. Prokhorov, *Sov. J. Quantum Electron. 10*:941 (1981).
9. S. M. Lee and S. M. Hauser, *Rev. Sci. Instrum. 35*:1679 (1964).
10. U. Kruger, R. Pepperl, and U. Schmidt, *Proc. IEEE 61*:992 (1973).

2

The Properties and Selection of Electro-Optic Materials

2.1 GENERAL

By its nature, the E-O effect in materials is very small. The change in optical susceptibility as a result of applied electric field is due to molecular, ionic, or electronic polarization. The polarization changes that can be achieved are small as the applied field is generally small in comparison with the field already existing within the material. For example, electrons move in a coulomb ($-e/r^2$) field which is typically $\sim 10^9$ V cm^{-1}. Although fields of up to $\sim 10^7$ V cm^{-1} have become available over the last decade or so in the beams of powerful lasers and have resulted in the observation of the Kerr effect at optic frequencies [1], fields of such magnitude are not available below infrared frequencies. For E-O devices operating in the dc to radiofrequency range, practical considerations limit the fields to $\sim 10^5$ V cm^{-1}. As a consequence, the refractive index changes due to electronic polarizability are exceedingly small. Even in the cases involving molecular or ionic polarizability, where the internal fields of the material are much weaker, the maximum refractive index change that can be achieved is generally only $\sim 10^{-3}$. Nevertheless, these small changes can be sufficient to allow practical E-O modulating or deflecting devices to be made providing the bulk material is of good optic quality.

Currently, the most useful E-O materials fall into two main groups. They are $NH_4H_2PO_4$ (ADP) and KH_2PO_4 (KDP) and its related isomorphs and ferroelectric materials related to perovskite ($CaTiO_3$). Other materials with more limited E-O applications include AB-type binary compounds, Kerr effect liquids, and ceramics in the (Pb,La) $(Zr,Ti)O_3$ system. The E-O properties of these materials are briefly reviewed here.

2.2 ADP, KDP, AND RELATED ISOMORPHS

Although it is now over 30 years since the E-O properties of ADP and KDP were first investigated [2-4], these materials remain the most widely used in E-O devices. This is mainly due to their fine optic quality. Good single crystals in these materials have been grown in sizes up to > 10 cm in diameter. They are grown at room temperature from a water solution and are free of the strain often found in other types grown at high temperatures. The growth process is slow, and high-quality large crystals can take up to 6 months to produce. Although the crystals are water soluble and fragile, they can be handled, cut, and polished without difficulty. At room temperature they are in an unpolarized phase and belong to the tetragonal crystal class 42 m. In the absence of an applied electric field, the crystals are uniaxial. The atoms, K, H, and P in KDP can be replaced by some of the atoms from corresponding columns in the Periodic Table without changing the crystal structure, for example, RbH_2PO_4 (RDP), KH_2AsO_4 (KDA), etc. The only nonvanishing E-O coefficients for this crystal class are $r_{41} = r_{52}$ and r_{63}.

ADP has an antiferroelectric transition at 148 K and KDP a ferroelectric transition at 123 K. In both materials the hydrogen can be replaced by its heavier isotope deuterium to form AD*P and KD*P, respectively. This has the effect of raising the transition temperatures of the materials to 242 and 222 K, respectively. In the ferroelectric material, the E-O coefficients, based on dielectric polarization, are the same temperature-independent constants for both KDP and KD*P. As a consequence, the higher transition temperature in KD*P leads to an increase in the room temperature linear E-O coefficients by a factor of $\sim$ 2.5. Since the dielectric constant in the ferroelectric materials closely follows a Curie-Weiss law, that is, $\varepsilon_r \simeq A/(T - T_c)$ where T_c is the transition temperature, the dielectric constant is also higher in KD*P. In fact, the E-O coefficient and dielectric constant increase in such a way that the quantity $r_{63}/(\varepsilon_r - 1)$ is roughly the same temperature-independent constant for all the KDP isomorphs despite the rapid increase in ε_r near T_c.

The refractive indices, and UV absorption which is associated with electronic transitions in the oxygen ions, are about the same for all materials in this group. The crystals are transparent down to $\sim$ 0.18 μm [5]. At the long-wavelength end, the infrared absorption is the result of O—H or O—D vibrations, and the cutoff frequency is approximately inversely proportional to the square root of the mass of the hydrogen isotope. Thus the low-frequency absorption edge for the deuterated salt occurs at roughly $\sqrt{2}$ times the wavelength for the undeuterated salts; for example, the cutoff frequencies for ADP, KDP, and KD*P are 1.4, $\sim$ 1.55, and 2.15 μm, respectively [6].

The resistivity of these materials is very high, typically $> 10^{10}$ Ω cm. In the visible, the optical loss is small and has been measured as only $\sim$ 0.5 dB m^{-1} in KDP at 632.8 nm, which is about as good as that found in the best fused quartz. The E-O coefficient r_{63} is practically independent of wavelength in the transparent region for both ADP and KDP. It is, of course, possible to obtain larger r_{63} by operating near the Curie temperature in KDP or its isomorphs, but both the loss tangent and dielectric constant increase and tend to infinity as T_c is approached. These two effects limit the high-frequency usefulness of the crystal near T_c. The former effect results in excessive heating, and the latter leads to a large drive current requirement.

2.3 FERROELECTRICS RELATED TO PEROVSKITE

A large class of ferroelectrics, having a structure of the form $A^{1+}B^{5+}O_3$ and $A^{2+}B^{4+}O_3$, are related to the mineral perovskite ($CaTiO_3$). These materials all have a centrosymmetric oxygen octahedron BO_6 as a central building block. They have crystal structures with different point symmetries that are derived from the perovskite structure by continuous lattice distortions. The cubic (paraelectric) form, which is usually the high-temperature phase, belongs to the nonpiezoelectric, nonferroelectric point group m3m. In the low-temperature phase, the most useful crystals are either tetragonal (4 mm), with the crystal axis along one of the original cube edges, or rhombohedral (3 m), with the crystal axis along one of the cube body diagonals.

Below the transition temperature (T_c), the linear E-O effect in the perovskites can be regarded as fundamentally a quadratic effect biased by the spontaneous polarization of the crystal; that is, polarization (P_k) can be written as

$$P_k = P_k^s + \delta P_k \tag{2.1}$$

in which P_k^s denotes the spontaneous polarization and

$$\delta P_k = \varepsilon_0(\varepsilon_k - 1)E_k \tag{2.2}$$

is the field-induced polarization. Consequently, using (1.21), we have

$$\Delta\left(\frac{1}{n^2}\right)_{ij} = g_{ijk\ell}P_k^s P_\ell^s + 2g_{ijk\ell}P_\ell^s \delta P_k \tag{2.3}$$

The first term in (2.3) describes the field-free birefringence of the crystal, while the second term describes the linear E-O effect. This equation with (2.2) shows that the linear E-O coefficient and the quadratic E-O coefficient are related by

$$r_{ij\ell} = 2g_{ijk\ell}P_k^s \varepsilon_0(\varepsilon_k - 1) \tag{2.4}$$

Above the transition temperature (in the paraelectric phase) a large dc field is often applied to the crystal to remove the center of symmetry and allow a linear E-O effect to be obtained. The field induces a large dc bias polarization P^{dc} so that a big refractive index change can be obtained by the use of a relatively modest modulating field. That is, above T_c, the induced polarization P^{dc} simply takes the role of P^s.

Generally, g_{mn} is about the same for all the perovskite-related ferroelectrics. Most reported measurements of g_{mn} have been obtained under conditions of constant stress, that is, at low frequencies. Consequently, as discussed in Chapter 1, they include a secondary contribution arising from electrostrictive strain. This contribution can be comparable in magnitude to the primary E-O effect if undamped acoustic resonances are allowed to occur. For frequencies substantially higher than the fundamental longitudinal acoustic mode of the crystal being used, the effect falls away rapidly.

All the perovskite-related ferroelectrics are insoluble in water. They are also more rugged and have larger refractive indices and dielectric constants than the E-O crystals in the former group. As a rule, they are transparent between ~ 0.4 and ~ 6.0 μm. The infrared absorption is caused largely by vibrations of the BO_6 octahedra and the UV absorption by electronic transitions in the oxygen ions. Currently, the most important materials are $KTa_{0.65}Nb_{0.35}O_3$ (KTN), $LiNbO_3$, $LiTaO_3$, and $BaTiO_3$.

KTN is a solid solution of two perovskites, $KTaO_3$ and $KNbO_3$, which have very nearly the same unit cell size (~ 4 Å) in their cubic phase but very different transition temperatures. These are ~ 4 and ~ 698 K, respectively. For the composition $KTa_{0.65}Nb_{0.35}O_3$, T_c is ~ 10°C; thus at room temperature KTN is just above the transition temperature and in its paraelectric phase. The properties of KTN and the other perovskites operated slightly above T_c are very temperature sensitive, since, as with the materials in the previous group, ε_r varies as $(T - T_c)^{-1}$. That is, T must be very carefully controlled. Even with samples of relatively high resistivity and low photoconductivity, the application of a large dc bias field leads to space charge effects which eventually reduce the internal biasing field.

If the temperature is reduced below T_c and the crystal poled into a single ferroelectric domain, the spontaneous polarization P^s removes the need for P^{dc}. Although P^s is usually several times greater than the induced P^{dc}, ε_r in the ferroelectric phase is much smaller than it is just above T_c. Well below T_c, r_{ij} and ε_r are insensitive to the temperature [7].

Barium titanate ($BaTiO_3$) was one of the earliest ferroelectric perovskites to be studied [8,9]. The crystals have a tetragonal (4 mm) phase between ~ 390 and 273 K and an orthorhombic phase below ~ 273 K. In the tetragonal phase the crystals can be poled into a single domain with the polar axis along one of the cube edges which exist above ~ 390 K. The only nonvanishing E-O coefficients are $r_{13} = r_{23}$, r_{33} and $r_{42} = r_{51}$. Both r_{13} and r_{33} are nearly temperature independent, but r_{42} increases rapidly as the tetragonal to orthorhombic phase transition is approached. At room temperature, r_{42}, at constant strain, is about 30 times bigger than r_{33}. Crystals with good optic and electrical properties with dimensions of $\gtrsim$ 1 cm can be grown routinely.

$LiNbO_3$ and $LiTaO_3$ are the most widely used of the perovskite group of ferroelectrics. Both are negative uniaxial crystals with rhombohedral (3 m) symmetry. They are a little unusual as ferroelectrics as their transition temperatures are so high. They are ~ 1470 and ~ 890 K, respectively. The direction of the spontaneous polarization in both crystals is parallel to the threefold rotation axis of the BO_6 block. The E-O tensor has the nonvanishing components $r_{13} = r_{23}$, r_{33}, $r_{22} = -r_{12} = -r_{61}$, and $r_{42} = r_{51}$. Crystals [10] of up to ~ 5 cm in diameter of good optic quality and high resistivity can be grown and if necessary poled into a single domain while near T_c by application of a small field (~ 1 V cm^{-1}). Because of the large T_c, considerable mechanical energy would be required to depole these crystals at room temperature. Hence, unlike both KTN and $BaTiO_3$, these crystals may be cut, polished, and roughly handled without creating additional domains. This ease of handling makes both $LiNbO_3$ and $LiTaO_3$ more widely used in practical (commercial) E-O devices than either KTN or $BaTiO_3$.

2.4 AB-TYPE BINARY COMPOUNDS

These compounds crystallize into either the cubic (43-m) zinc blende structure at room temperature or the hexagonal (6-mm) wurzite structure at higher temperatures. In the cubic phase there is only one E-O coefficient, that is, $r_{41} = r_{52} = r_{63}$. In the hexagonal phase there are three coefficients: $r_{13} = r_{23}$, r_{33} and $r_{54} = r_{42}$.

The main interest in these materials has been for E-O devices in the infrared, particularly at 10.6 μm for use with the technologically important CO_2 laser. GaAs, ZnTe, ZnS, CdS, and CdTe are among the most widely used materials in this group that transmit out beyond 10 μm and are available as single crystals with dimensions $\gtrsim$ 1 cm. Although the E-O coefficients of these materials are small, typically $\lesssim$ 10% of the values for the crystals in the previous groups, their refractive indices are high (~ 2 to 4) so that the important parameter n^3r from (1.16) is comparable.

The fact that the 43 m crystals are cubic and hence optically isotropic makes them potentially attractive for light beam deflection applications because they have a large acceptance angle. The hexagonal crystals are birefringent and consequently, from this point of view, are not so attractive.

The group covers a wide spectrum of materials with widely different optic and mechanical properties [11].

2.5 KERR EFFECT LIQUIDS

The advantages of using a fluid (potentially the ideal homogeneous medium) in E-O deflectors has led to renewed interest in Kerr effect

liquids. There have been recent extensive studies of the behavior of nitrobenzene Kerr cells [12,13] and attempts to find Kerr liquids more suitable than nitrobenzene [14].

For the simple case of a polarization switch providing $\pi/2$ rotation at visible frequencies, a Kerr cell $\sim$ 1 cm long using nitrobenzene requires a field of $\sim$ 35 kV cm^{-1}. Such high fields can lead to heating and optic inhomogeneity in the media and, in extreme cases, to electrical breakdown. These effects can be prevented only if the solution is extremely pure. For high-field dc operation an ultralow charge carrier concentration (below $\sim 2.10^{11}$ cm^{-3}) is required. Larger residual currents lead to space charge, due to polarization effects at the electrodes, which lowers and makes inhomogeneous the field distribution in the bulk material.

Electrochemical processes at the electrodes and hydrodynamic current caused by injected charge carriers also contribute strongly to the residual current. Further, the accelerated charge carriers cause turbulence, which tends to disorientate the alignment of the molecules. Filippini [12] has shown that these effects can be mitigated to some extent by the use of electrodialytic electrodes. For pulsed high fields of duration less than $\sim 10^{-4}$ sec, the conductivity of nominally pure nitrobenzene is not usually a problem, but it can prohibit the cell being dc biased to reduce the modulating (signal) power requirement.

Although there are theoretical models that allow Kerr constants to be estimated for gases, there is no satisfactory means of quantitatively predicting the values for liquids, because the internal field in a liquid deviates strongly from the external applied field.

Recent studies of a variety of polar substances by Blanchet [15] and Kruger et al. [14] have not revealed any with appreciably superior properties to nitrobenzene for use in Kerr cells. As a consequence of this and of the problems outlined above, and the difficulty of purifying highly polar molecules, Kerr effect liquids remain an unattractive option for many E-O applications.

2.6 ELECTRO-OPTIC CERAMICS IN THE $(Pb,La)(Zr,Ti)O_3$ SYSTEM

Lanthanum-modified lead zirconate titanate (PLZT) ceramic materials have been investigated since 1969 for their electro-optic properties, and by controlling the La/Zr/Ti ratio both the linear and quadratic [16,17] electro-optic effects have been obtained. PLZT is ferroelectric when exhibiting the linear and quadratic electro-optic effect and Δn follows a hysteresis loop with applied electric field, although this is less apparent in the quadratic cubic phase. It has so far proved difficult to produce material of good quality and experimenters have resorted to the use of thin plates in a longitudinal quadratic electro-optic modulation mode. Large apertures have been obtained [18] using interdigital electrodes for light modulation applications such as eye protective goggles, photographic shutters and spatial light modulators [19]. Although the refractive index change can be large (e.g., $\sim 0.5 \times 10^{-8}$ m/V for the quadratic material), the response time is generally slow because of the ferroelectric domain orientation processes. The fastest switching time reported to data [18] for an experimental device has been 10 μsec and this in particular makes the material unattractive for most electro-optic beam deflection applications.

2.7 MATERIAL SELECTION

Currently, there are only a handful of materials that are suitable for use in E-O deflectors. The requirements are stringent, and many are common to all devices. Material requirements generally include high resistivity ($\gtrsim 10^8$ Ω cm), good homogeneity (refractive index variations of less than $\sim 10^{-6}$), and a large E-O effect ($\Delta n \gtrsim 10^{-4}$).

There are further material requirements for devices operated at high frequency. These often come down to minimizing the effects of heating. Crucial parameters include (1) the dielectric loss, which must be small to minimize heating; (2) the temperature dependence of birefringence, which must also be small for good stability; and (3) the heat conductivity, which must be large to reduce thermal gradients. Often there is a *figure of merit* appropriate to a material when used in a particular device. This figure of merit is used to weight the various material parameters to reflect their relative importance for a particular application and allow selection of the most suitable material available. Where such figures of merit exist, they will be discussed in subsequent chapters along with the particular deflector designs and applications for which they are relevant.

REFERENCES

1. M. A. Duguay and J. W. Hansen, *Appl. Phys. Lett.* *15*:192 (1969).
2. B. H. Billings, *J. Opt. Soc. Am.* *39*:797 (1949).
3. B. Zwicker and P. Scherrer, *Helv. Phys. Acta* *16*:214 (1943).
4. R. Carpenter, *J. Opt. Soc. Am.* *40*:225 (1950).
5. W. L. Smith, *Appl. Opt.* *16*:1798 (1977).
6. I. P. Kaminow and E. H. Turner, *Proc. IEEE* *54*:1374 (1966).
7. W. Haas and R. Johannes, *Appl. Opt.* *6*:2007 (1967).
8. A. R. Johnson and J. M. Weingart, *J. Opt. Soc. Am.* *55*:828 (1965).
9. A. R. Johnson, *Appl. Phys. Lett.* *7*:195 (1965).
10. K. Nassau, H. J. Levinstein, and G. M. Loiacono, *J. Phys. Chem. Solids* *27*:989 (1966).
11. R. J. Pressley, ed., *Handbook of Lasers*, The Chemical Rubber Co., Cleveland (1971).

12. J. C. Filippini, *J. Phys. D. 8*:201 (1975).
13. H. Krause and K. Barner, *J. Phys. D. 10*:2429 (1977).
14. U. Kruger, R. Perrerl, and U. Schmidt, *Proc. IEEE 61*:992 (1973).
15. M. Blanchet, in *Proc. 8th Int. Symp. on High Speed Photography, Stockholm, Sweden* (1968), p. 64.
16. G. H. Haertling, *J. Am. Ceram. Soc. 3*:269 (1972).
17. G. H. Haertling and C. E. Land, *IEEE Trans. Ultrason. 3*:269 (1972).
18. T. J. Cutchen, J. O'Harris, and G. R. Laguna, *Appl. Opt. 14*:1866 (1975).
19. K. Ueno and T. Saku, *Appl. Opt. 19*:164 (1980).

3

Principles of Electro-Optic Deflectors

3.1 DIGITAL LIGHT DEFLECTORS (DLDs)

3.1.1 General

DLDs have discrete deflection positions. The basic deflection cell consists of a polarization modulation element (E-O switch) in conjunction with a birefringent discriminator (polarizer). In operation, the switch can be activated to change the polarization state of an incoming light beam. This results in two possible beam directions from the polarizer corresponding to the ordinary (o) and extraordinary (e) ray directions. Depending on the type of polarizer used, the two beams can be either angularly or linearly displaced.

DLDs possess the following important features:

1. N stages can be cascaded to produce 2^N discrete beam positions.
2. They require only a two-state control voltage and as a consequence can convert binary coded digital voltage pulses directly into optic displacement.
3. The addressable beam directions are independent of fluctuations in control voltage because the directions are established by the polarizers; that is, steering and switching functions are separated.
4. Any beam position can be obtained at random as all switches can be addressed simultaneously. There is no *flyback* as on a CRT (cathode ray tube).
5. A two-dimensional displacement can be achieved with two sets of deflection stages with mutually perpendicular beam displacements.

3.1.2 Digital Deflector Unit

Polarizers

The basic digital deflector unit consists of a polarization discriminator and an E-O switch. Several types of polarization discriminator have been suggested and used in DLD systems. They include a split-angle birefringent plate, a total internal reflection polarizer, a Wollaston prism, and a simple birefringent wedge. These polarizers are shown in Fig. 3.1.

Split-angle birefringent plate This polarizer is a birefringent plate orientated so that the o-ray passes through undeviated, whereas the

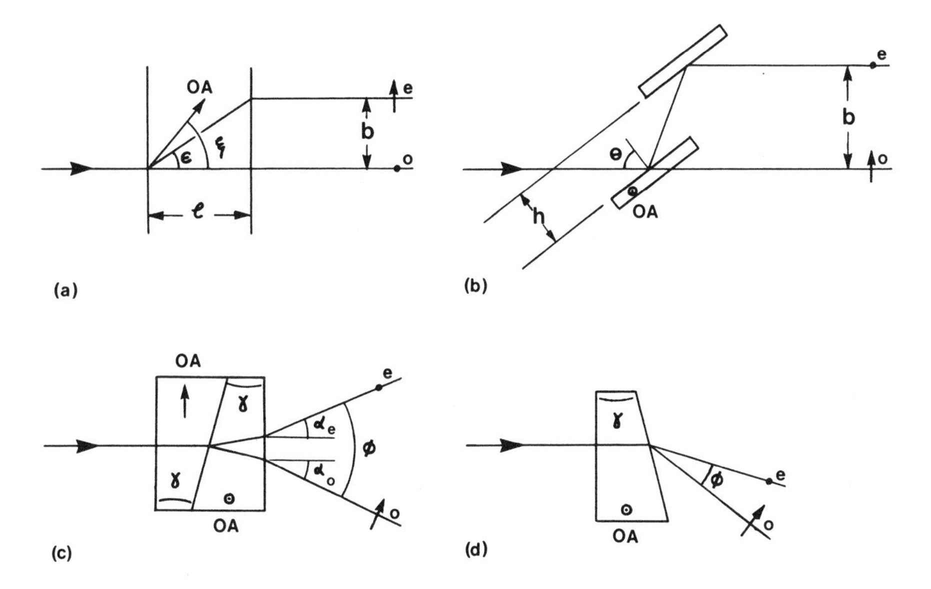

Figure 3.1 Four polarization elements proposed or used in DLD systems. (a) Split-angle birefringent plate, (b) total internal reflection polarizer, (c) Wollaston prism, and (d) simple birefringent wedge. Polarization directions (E vector) are indicated in the figure for the case of a negative uniaxial material, for example, calcite.

e-ray is refracted and leaves the element displaced and parallel with the o-ray. With the notation used in Fig. 3.1 it can be seen that the separation (b) of the two possible output beams is $\ell \tan \varepsilon$. It can be shown [1,2] that the maximum value of ε is given by

$$\varepsilon_{max} = \arctan \frac{n_o^2 - n_e^2}{2n_o n_e} \qquad (3.1)$$

where n_o and n_e are the refractive indices in the birefringent plate for the o- and e-rays, respectively. The values of ε_{max} are fairly modest. For example ε_{max} is ~ 5.9° in calcite and ~ 9.17° in sodium nitrite [3] at 632.8 nm. The corresponding value of ξ for the orientation of the optic axis is

$$\xi_{max} = \arctan \frac{n_e}{n_o} \qquad (3.2)$$

For a noncollimated beam, the extraordinary output component suffers from aberrations in passing through the polarizer. Kulcke et al. [2] have shown that there is an OA orientation (other than ξ_{max}) that minimizes this aberration.

Total internal reflection (TIR) polarizer In this case the polarizer consists of a thin birefringent plate immersed in an optic-quality oil of refractive index (n_m) chosen to match that of the higher of the two indices of the plate. Total internal reflection occurs when the input beam angle (θ) is equal to or larger than the critical angle θ_c, where

$$\sin \theta_c = \frac{n_e}{n_m} \quad \text{for } n_e < n_o = n_m \tag{3.3}$$

A second plate, which is either a mirror or an optic element with an index of refraction $n' < n_e$, reflects the TIR beam parallel to the transmitted one. The beam path displacement (b) is $2h \sin \theta$, and the path length difference between the polarization components is $n_m b \tan \theta$.

Wollaston prism A Wollaston prism consists of two birefringent wedges cemented together but with their respective OAs orthogonal. As a result an o-ray in the first prism becomes an e-ray in the second and vice versa. This leads to different angles of refraction at the interface for the two input polarizations. For normal incidence the angle of divergence ϕ between the two exiting components of the beam from the polarizer is given by [4]

$$\phi \simeq \alpha_o + \alpha_e \simeq 2(n_e - n_o) \tan \gamma \tag{3.4}$$

For angles of incidence $\alpha_i \simeq 0$, the angle ϕ is very nearly the same for small prism angles (γ), but the two angles α_o and α_e change to $\alpha_o \pm \alpha_i$ and $\alpha_e \pm \alpha_i$, respectively. In a cascaded DLD system these output angles are the angles of incidence for the following deflection stage.

Simple birefringent wedge A birefringent prism that introduces a small angular separation between the o and e beam components can be used as a polarization discriminator. For a right-angle prism of apex angle γ, the beams are separated by ϕ, where

$$\phi \simeq (n_o - n_e) \sin \gamma \tag{3.5}$$

That is, in the small-angle limit the angular separation is half that of the Wollaston prism. The simple wedge polarizer has the advantage that it reduces the volume requirement of birefringent material. This is particularly important from a cost point of view in the construction of large-aperture deflection cells. A disadvantage is that the output beam components are not deviated symmetrically around the line of the input beam. If required, this can be corrected for by using a glass compensating wedge of refractive index intermediate between that of n_e and n_o of the polarizer [5].

All the polarization discriminators above introduce, to a lesser or greater extent, optic path length differences for the o and e polarization components. Consequently, if the beam in a DLD system is focused, there is a shift in the focal plane for each beam position, the magnitude of which depends on the beam path. Techniques of path length compensation exist so that the source point is projected as an equal-sized image at every output point [2].

If a polarized beam entering a birefringent crystal is not fully collimated, the beam is split into two components. The major component propagates through the crystal according to the polarization direction of the incident beam, whereas the minor part propagates as the complementary beam and appears as background light in the unwanted position, that is, as *cross-talk*. Figure 3.2 is an example of the dependence of this cross-talk on convergence angle (β') for an optimally orientated calcite split-angle polarizer used at 632.8 nm. The figure is taken from the paper by Kulcke and co-workers [2]. For most applications deflectors are designed around a maximum cross-talk level of $\sim 1\%$ per stage. From Fig. 3.2 we see that this restricts a calcite split-angle polarizer to use with a beam of convergence angle less than $\sim 2.7°$.

The calculations used for cross talk by Kulke and co-workers are given in an earlier paper [6] for a light beam of oblique incidence on a z-cut calcite plate. In this analysis the coordinate system of the incident light beam is given by x'y'z', where x' is parallel to the x axis of the crystal surface and ν is the angle between the surface normal and the z' axis of the light beam. For light polarized along the y' axis the fractional intensity in the wrong channel is given by

$$\mathrm{Ix}(\beta') = \frac{1}{2\pi\beta'} \int_0^{2\pi} \int_0^{\beta'} \sin^2 \zeta \, d\eta \, d\beta \tag{3.6}$$

The angular coordinates $\beta\eta$ form a principal plane which intersects the circular cut of the index ellipsoid of the birefringent crystal under an angle ζ relative to the x axis and

$$\mathrm{Tan}\,\zeta = \frac{-\sin\beta \, \cos\eta}{\sin\beta \, \sin\eta \, \cos\nu + \cos\beta \, \sin\nu}$$

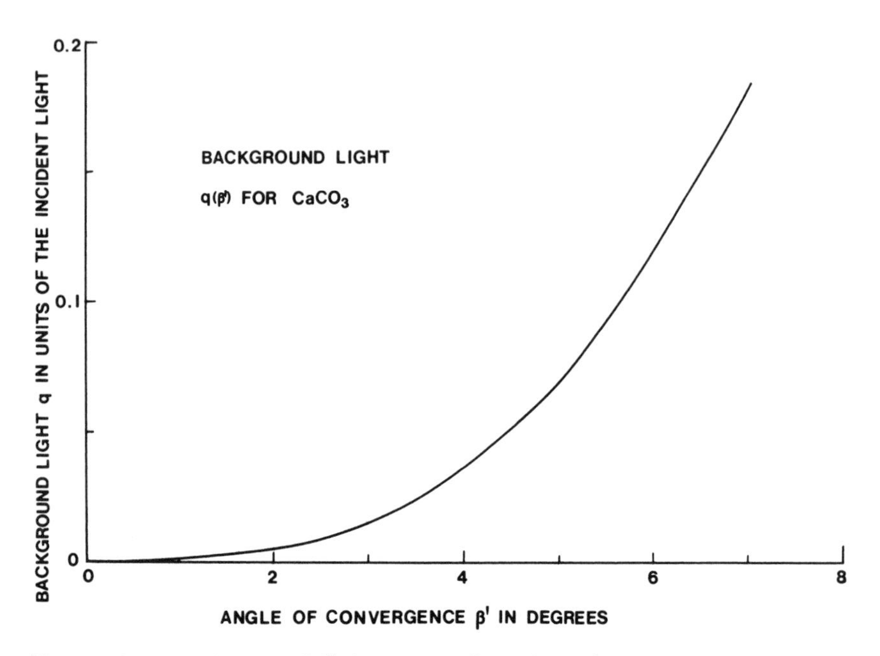

Figure 3.2 Background light q as a function of the maximum convergence angle β' (β' equals light cone semiangle) at 632.8 nm in a calcite split-angle polarizer at optimum orientation of the OA. (From Ref. 2.)

Electro-optic Switch

The number of E-O materials that can be used in practical DLD systems for the switch element is severely limited. Some Kerr liquids, for example, nitrobenzene, and some linear E-O crystals, such as KD*P, have been used. As with the polarizers discussed above, the natural birefringence of the latter class of materials can give rise to cross-talk when used with an angled or convergent beam. A second cause of cross-talk, common to both isotropic and anisotropic switch materials, is error in the control voltage.

For XDP Z-cut linear E-O crystals, Ley et al. [7] have shown that for small input angles (ξ) to the Z crystal axis the background light approximates to

$$\frac{\Delta I}{I} = \left[\frac{\pi \ell}{2\lambda} \; \frac{n_o^2(n_o^2 - n_e^2)}{n_e^2} \, \xi^2\right]^2 \tag{3.7}$$

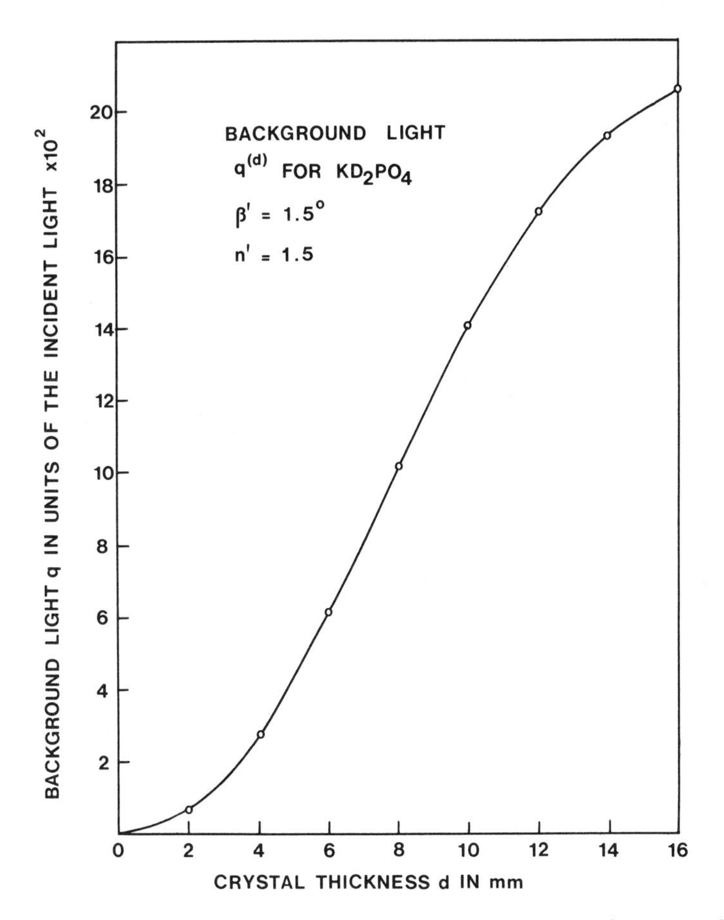

Figure 3.3 Background light q due to a beam with a maximum convergence angle $\beta' = 1.5°$ passing through a KD*P crystal immersed in a liquid with a refractive index $n' = 1.5$. (From Ref. 2.)

Here ℓ is the crystal thickness (the other symbols have their usual meaning).

Mechanical strength usually limits the aspect ratio of E-O crystals to less than $\backsim$ 10:1; for example, a 25-mm-aperture crystal will have a minimum thickness of $\backsim$ 2.5 mm. Using the data for KD*P in Table 1.1, for a crystal of this thickness, Eq. (3.7) shows that the beam angle ξ must be $\leqslant 0.8°$ for the cross-talk to be $\leqslant 1\%$.

Kulcke et al. [2] have made a similar calculation but for a convergent beam directed along the Z crystal axis. In this case the cross-talk contributions of all the rays at angles ξ, up to the maximum beam convergence angle β', need to be summed. The results of the calculations made by Kulcke and co-workers for KD*P are shown in Fig. 3.3 for the particular case where $\beta' = 1.5°$. It can be seen that 1.5° is about the maximum convergence angle that can be tolerated if the cross-talk is to be < 1% for a 2.5-mm-thick crystal. This is a more stringent collimation requirement than that for the polarizers.

The cross-talk as a function of control voltage error ($\Delta V/V_{\lambda/2}$) for a DLD with a linear E-O switch is given by

$$\frac{\Delta I}{I} = \sin^2\left(\frac{\pi}{2}\frac{\Delta V}{V_{\lambda/2}}\right) \simeq \left(\frac{\pi}{2}\frac{\Delta V}{V_{\lambda/2}}\right)^2 \tag{3.8}$$

That is, the voltage error must be less than ~ 6% for the beam cross-talk not to exceed 1%. This is not a very stringent requirement. For the case of a material exhibiting the quadratic E-O effect the requirement is even more relaxed; that is,

$$\frac{\Delta I}{I} = \sin^2\left[\frac{\pi}{2}\left(\frac{\Delta V}{V_{\lambda/2}}\right)^2\right] \simeq \left[\frac{\pi}{2}\left(\frac{\Delta V}{V_{\lambda/2}}\right)^2\right]^2 \tag{3.9}$$

For $\Delta I/I \simeq 1\%$, the tolerable voltage error is ~ 25%.

Quadratic E-O materials have the further advantage that they do not exhibit significant piezoelectric effects. These can be quite large in linear E-O materials and lead to cross-talk at high switching frequencies.

3.1.3 DLD Arrays

Resolution Limit

In a large DLD system with Wollaston- or calcite-type wedge polarizers the angle ξ between the beam and the switch (e.g., Pockels cell) optic axis increases by a factor of 2 per stage in order to conserve angular resolution. Consequently, cross-talk generally sets an upper limit to the number of stages that can be used.

For an $n \times m$ DLD array (i.e., one with n elements giving horizontal deflection and m giving vertical deflection), $2^n \times 2^m$ accessible output positions are achieved. Assuming that they are equally separated, the nth displacement b_n is given by

$$b_n = 2^{n-1} b_1 \tag{3.10}$$

where b_1 is the displacement obtained in the stage producing the smallest deflection. From an availability and cost point of view it is important to employ the smallest possible crystals in the deflector system. Therefore, the output spot size and spacing are generally chosen to be of the order of the resolution limit obtainable with a convergent light beam. This requires the beam to be focused on the output face of the last deflector element. Optical relaying is then used for display purposes. Assuming that 1.5° is the maximum beam convergence angle that can be tolerated in the DLD system, the diffraction-limited focal spot diameter (which sets the resolution limit for a perfect DLD system) is $\sim 30\ \mu m$ at 632.8 nm. For a deflector system with an output aperture limited to 30 mm, Eq. (3.10) gives 11 as the maximum value of n; that is, a two-dimensional DLD array of this aperture can have a maximum of $2^{11} \times 2^{11} > 10^6$ resolved beam positions.

Optic Performance

To achieve uniform electric fields in DLDs employing large-aspect-ratio Z-cut crystals, resistive electrodes can be applied to the faces through which the optical beam passes. Typically, electrodes of resistivity $\sim 500\ \Omega$ square are used. The optical losses that result can be reduced by applying antireflection coatings to the surfaces and by immersion of the crystals in an index-matching fluid. These techniques can result in practically the entire transmission loss of a DLD system being due to electrode absorption. An n-stage system having 2n electrodes will have an overall transmission limited to t^{2n}, where t is the transmission of each electrode. Clearly, a DLD system of this design with a large number of deflection stages will have low transmission even if the electrode absorption is only a few percent.

At low switching frequencies the fraction of the transmitted beam that can be focused within a diffraction-limited spot size depends on the quality of the optical components that comprise the DLD system. Kruger et al. [8] have analyzed the case where an input beam of Gaussian spatial profile is used with a deflector system that suffers from a Gaussian distribution of refractive index inhomogeneities which are small in size compared to the beam diameter and are evenly distributed across the aperture. They assumed that the index fluctuations were statistically independent of each other. For N inhomogeneities per square centimeter of average cross-section $\sigma\ cm^2$, they found that $\sqrt{N}\,\sigma$ was the mean phase variation impressed on the beam and that the output intensity was reduced by a factor $\exp(-N\sigma^2)$. The "lost" radiation was found to be scattered rather broadly. As an example, they calculated that 10% of the input beam energy was scattered if the refractive index variation in a 20-stage DLD system distorted the beam wavefront by more than $\lambda/70$ per stage. For a 50% overall scattering loss, this figure reduces to $\lambda/30$ per stage [9]. These results imply that the very finest optic-quality components must be used in a large DLD system.

High-Frequency Operation

The performance of DLDs at high frequency can be limited by several factors. These include the geometry, heating (i.e., tan δ losses in the E-O material and resistive effects at the electrodes), and the onset of piezoelectric resonances.

Geometry The aperture limitation of DLD systems requires the total length of the device to be restricted in order to prevent beam *walk-off* and the outermost beam positions losing intensity. Based on the assumption that a loss of 1% is tolerable for these beam positions, Kurtz [10] has derived an expression for the limitation of the capacity-speed product (Rf) of a DLD system using KD*P E-O switches. Here R is the number of light beam directions in one dimension (linear capacity of the DLD) and f is the fundamental frequency of the square-wave voltage by which the polarization switch is operated and which corresponds to a switching frequency of 2f light beam positions per second. Kurtz obtained

$$\mathrm{Rf} \leqslant \frac{2 \times 10^{-2}}{\sqrt{3}} \frac{\ell P_r}{\gamma_f \beta \ell_m \lambda V_{\lambda/2}^2 \varepsilon_0 \varepsilon_z} \tag{3.11}$$

where β is the ratio of the basic angular unit of deflection to the diffraction angle (ideally $\beta \sim 1$), ℓ_m the length of a deflection stage, P_r the reactive drive power for a polarization switch, and ℓ the thickness of the KD*P crystal plate of relative dielectric constant ε_z. (γ_f is a numerical factor, which for square-wave operation is ~ 2.4.) Application of this expression to a typical example, that is, $P_r \sim 200$ W, $\ell \sim 1$ mm, $\ell_m \sim 7$ mm, $\lambda = 632.8$ nm, $\beta = 1$, and $\varepsilon_z = 50$, gives

$$\mathrm{Rf} < 2^8 \times 10^4 \ \mathrm{sec}^{-1} \tag{3.12}$$

That is, an eight-stage DLD system using KD*P switches is limited by its geometry to a maximum operating frequency of ~ 10 kHz.

Heating The capacity-speed product given by (3.11) can only be calculated assuming a value for the maximum permissible reactive drive power P_r to each stage of the DLD. The maximum of P_r is determined by the onset of unacceptable heating and consequent strain effects in the E-O switch material. This limit can be set either by the tan δ losses in the bulk material or by heat generated in resistive electrodes if they are used. For thin crystals in which the temperature gradient is only radially (x-y plane) significant, Pepperl [5] has calculated the maximum tolerable reactive drive power following a model due to Kaminow [11]. For the case of KD*P, a depolarization of ~ 1% resulting from shear strain occurs if the unit volume power dissipation exceeds ~ 0.27 W. Since tan $\delta \sim 10^{-3}$ for KD*P, this corresponds to a reactive power of $0.27/\tan\delta \sim 270$ W.

The resistivity of the electrodes is a second source of heat in switches using linear Z-cut E-O crystals. The power dissipated in this case is given by

$$P = \frac{1}{2} \omega^2 C^2 V_{app}^2 R_e \tag{3.13}$$

where R_e is the square resistance of both electrodes, C is the cell capacitance, and V_{app} the applied voltage of angular frequency $\omega = 2\pi f$. Rearranging (3.13) yields

$$f = \frac{1}{2\pi C V_{app}} \sqrt{\frac{2P}{R_e}} \tag{3.14}$$

A reasonable criterion that can be used in setting the upper limit for the acceptable power dissipation in the electrodes is that it should not exceed that due to the tan δ losses in the bulk material. For the example above this gives $P < 4$ mW. Assuming that $C \sim 100$ pF, $V_{app} \sim 2$ kV, and $R_e \sim 500\ \Omega$ are values typical for a switch, (3.14) gives $f_{max} \simeq 4$ kHz, that is, a frequency limit similar to that imposed by the tan δ losses.

Piezoelectric effects A further constraint on the high-frequency operation of switches using linear E-O materials is their piezoelectric nature. DLD systems with linear E-O switching elements must be operated at low frequencies under the condition of constant stress so that piezoelectric resonances are avoided. Ley et al. [7] have shown that the typical strain relaxation times for ADP and KD*P crystals of order ~ 1 cm in dimension are in the range 1 to 10 μsec. Consequently resonances are avoided by switching at frequencies below ~ 10 kHz.

The preceding discussion has mainly concerned switches employing linear E-O, Z-cut crystals. Specifically, KD*P has been discussed, as this is a widely used material. Despite this, from a speed and heating point of view, Kerr cells are considerably more attractive; for example, nitrobenzene has a low ε and does not exhibit piezoelectric effects so that access times $< 0.5\ \mu$sec are obtainable.

Optimizing DLD Design

The capacity-speed product (CSP) which is commonly used as a figure of merit for DLD devices has been defined by (3.11) for the maximum linear resolution (R) and maximum deflection rate (f) on the assumption that the deflection rate is equal for each stage of the DLD system. Schmidt et al. [9] have considered the case where the system geometry is optimized for minimum drive power to the last deflector stage (N). Their analysis is for switches driven in the transverse mode so that it is applicable to both linear and quadratic E-O materials. For minimum drive power the relations linking the height (d), width (w), and length (ℓ) of the last deflector were found to be

$$d_N = 2.82\sqrt{R\beta\ell\lambda} \tag{3.15}$$

and

$$w_N = 3.53\sqrt{R\beta\ell\lambda} \tag{3.16}$$

The corresponding capacity-speed product for a DLD system with linear switches driven in the transverse mode configuration was found to be

$$Rf = \frac{P}{4.95\varepsilon\varepsilon_0 V_{\lambda/2}\beta\lambda} \tag{3.17}$$

In deriving this last expression it was assumed that the DLD comprised an equal number of x and y stages and that the beam-splitting angle increased in the direction of beam propagation.

For liquid E-O materials that always exhibit the quadratic effect, $V_{\lambda/2}$ in (3.17) should be replaced by $B/2\ell$, where B is Kerr's constant defined by (1.28). In this case a bias voltage can be applied to the polarization switch so that it operates between the mth and (m + 1)th $\lambda/2$ wave points. Then the CSP becomes [9]

$$Rf = P(\sqrt{m+1} - \sqrt{m})^{-2}\,\frac{2B}{\varepsilon\varepsilon_0}\,\frac{1}{4.95\beta\lambda\ell} \tag{3.18}$$

The advantage of applying a bias voltage is apparent if one considers the simplest case where m = 1. This reduces the power required to drive the switch by a factor of 6.

In the limit where m >> 1, (3.18) becomes

$$Rf = P\,\frac{(4B)^2}{\varepsilon\varepsilon_0\beta\lambda}\,\frac{V_m^2}{d^2}\,\frac{1}{4.95} \tag{3.19}$$

Here V_m denotes the applied voltage required for an $m\lambda/2$ waves phase difference.

For solid-state quadratic materials with a bias polarization P_b and E-O coefficient g_{ij}, (3.19) becomes

$$Rf = P\left(\frac{4a}{\lambda}P_b\right)^2\,\frac{\varepsilon\varepsilon_0}{\beta\lambda}\,\frac{1}{4.95} \tag{3.20}$$

where, for example, $a = \frac{n^3}{2}(g_{11} - g_{12})$ if the polarization is in the x direction.

For display applications, the stages of small splitting angles in a DLD usually have to be switched at higher rates than those of large splitting angles. In this case Schmidt et al. [9] suggest that it might be of greater importance to design the deflector for higher resolution rather than minimum drive power in the last stage. An optimization of the geometry from this point of view for a two-dimensional deflector led Schmidt et al. to (1) a stage configuration where the y stages precede the x stages and (2) a capacity factor for E-O liquids of

$$R = \frac{V_{app}^2}{2} \frac{B}{\beta\lambda} \tag{3.21}$$

A comparison of the relative CSPs of solid-state DLD systems with performance limited by tan δ heating has been carried out by Kruger et al. [8]. In a DLD system with $R = 2^{10}$ (i.e., 10 stages), they calculated that KTN or $LiNbO_3$ electro-optic switches could be operated at rates of $\sim 10^6$ addresses per second and $LiTaO_3$ at $\sim 10^5$ addresses per second. In contrast the high drive voltage required for KD*P electro-optic switches leads to high dissipated power and (as we have seen above) maximum switching rates restricted to the range $< 10^4$/sec.

Finally, since the CSP is proportional to $1/\lambda V_{\lambda/2}^2$, it is worth noting that there is a considerable advantage in operating near the short-wavelength cutoff of the DLD optic components. In this case, the minimum laser wavelength and beam power are limited by absorption, causing bulk heating. For a DLD system using nitrobenzene Kerr cells and an He-Ne laser operating at a wavelength of 632.8 nm, this heating effect limits the beam power to $\sim$ 70 to 100 mW [8].

3.2 ANALOG LIGHT DEFLECTORS

3.2.1 Prism and Multiple-Prism Deflectors

Basic Prism Unit

The angle (ϕ) through which a prism deflects an optic beam is a function of its refractive index (n). Consequently, a change in n, due to the E-O effect, leads to a change in ϕ. This is the basis of the E-O prism deflector.

The deflection produced by a prism can be calculated by applying Snell's law at the boundaries. The change $\Delta\phi$ in the deflection angle produced by a change Δn in refractive index is given to first order by carrying out the proper differentiation of the resultant equation and yields [12]

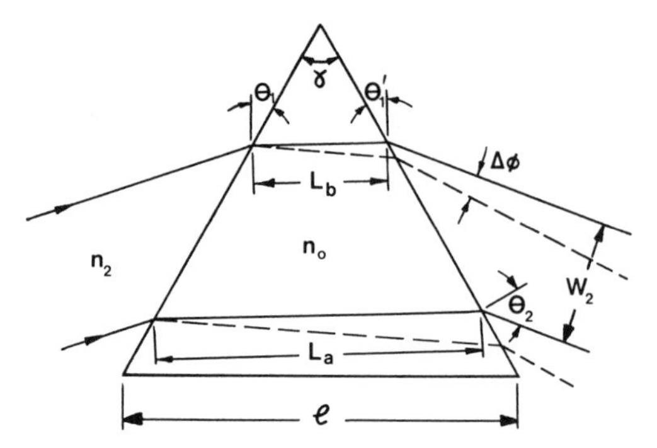

Figure 3.4 Beam deflection by an E-O prism element. (From Ref. 12.)

$$\Delta\phi = \frac{\Delta n(\tan\theta_1 \cos\theta_1' + \sin\theta_1)}{n_2 \cos\theta_2} \tag{3.22}$$

$$= \frac{\Delta n(L_a - L_b)}{n_2 W_2} \tag{3.23}$$

Here n_2 is the refractive index of the medium in which the prism is immersed, and the angles θ_1, θ_2, and θ_1' and the distances L_a, L_b, and W_2 are defined in Fig. 3.4. It is usual to express the deflection $\Delta\phi$ in terms of *number of resolvable spots* (N_R). For the ideal case of a beam with diffraction-limited divergence (θ_R), the Rayleigh criterion gives

$$\theta_R = \frac{\varepsilon\lambda}{n_2 W_2} \tag{3.24}$$

where λ is the free-space optic wavelength and ε is a factor ($\gtrsim 1$) that depends on the beam spatial intensity distribution. Consequently, N_R is given by

$$N_R = \frac{\Delta\phi}{\theta_R} = \frac{\Delta n(L_a - L_b)}{\varepsilon\lambda} \tag{3.25}$$

We see that under the most favorable conditions ($\varepsilon = 1$), N_R is numerically equal to the induced optic path difference for the marginal rays of the beam in the prism, expressed in number of waves.

It can be shown that N_R is maximized by using the prism at minimum deviation and at full aperture [13]. In this case, $L_a - L_b$ equals the prism base length (ℓ), and N_R is independent of the prism apex angle (γ). To minimize the E-O material requirement, γ should be large but kept below $2 \sin^{-1}(n_2/n)$ to avoid TIR losses.

The deflection changes provided by a single prism of E-O material are generally rather modest. For example, in a typical case $\Delta n \simeq 10^{-4}$, $\ell \simeq 1$ cm, $\lambda \simeq 0.5$ μm, and (3.25) gives $N_R \simeq 2$. Consequently, for practical deflectors, multiple-prism devices are generally constructed.

Multiple-Prism Deflector

Equation (3.25) shows that the number of resolved beam positions from a prism deflector is proportional to the induced optic path difference for the marginal rays of the beam. This quantity can be increased by a series arrangement of E-O detector prisms. By alternating the crystal orientation or field in consecutive prisms, the zero field deflection of the device can be eliminated. A schematic of such an arrangement is shown in Fig. 3.5. In this case, the total internal accumulated deflection can be regarded as the sum of the separate deflections resulting from the upright prisms of refractive index $n_0 + \Delta n$ that are located in a medium of index $n_0 - \Delta n$. The smallness of the E-O deflection usually allows the beam to be treated as near parallel to the base, and hence the minimum deviation condition applies in each prism. As a result, $\Delta\phi$ is given by [14]

$$\Delta\phi = \frac{2L\Delta n}{Wn_2} \tag{3.26}$$

where L is the deflector length and W the aperture. Consequently, the maximum number of resolved beam positions N_R is

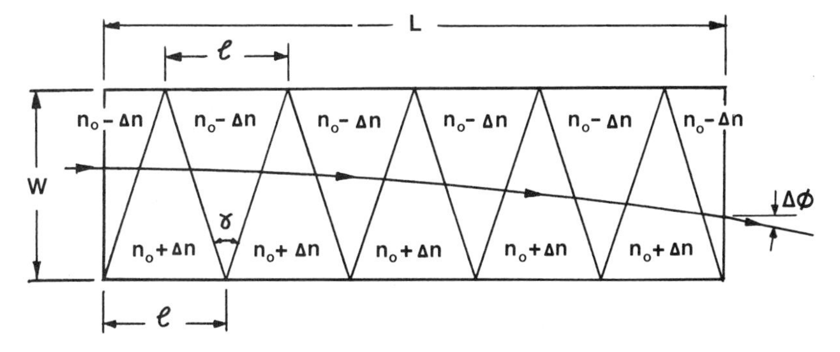

Figure 3.5 Multiple-prism deflector comprising a series arrangement of E-O prism elements.

$$N_R = \frac{\Delta\phi}{\theta_R} = \frac{2L\Delta n}{\varepsilon\lambda} \tag{3.27}$$

That is, in a cascaded deflector, N_R is increased by a factor $2L/\ell$ over that for a single-prism device.

Optimized Design

The limit to the number of prisms that can be cascaded in a series deflector is set by the criterion that the beam must not hit the edge of the exit aperture of the deflector. As with DLDs, this requirement naturally leads to the use of weakly focused optic beams. For the case of a deflector in which the total length of the light path is L and which has mirror symmetry about L/2, the rays of the deflected beam, when extended backwards, intersect with the undeflected rays at the L/2 plane. This is indicated schematically in Fig. 3.6, where, for convenience, the deflector has been considered to be immersed in an index-matching medium. With the beam focused at a distance L_1 beyond the exit aperture of the deflector, the deflection distance is $(L/2 + L_1)\Delta\phi$, and the resolved spot size is $\theta_R(L + L_1)$. Consequently, for a deflector used with a focused beam, N_R is given by

$$N_R = \frac{\Delta\phi}{\theta_R}\left[1 - \frac{L}{2(L + L_1)}\right] \tag{3.28}$$

That is, as a result of focusing the beam to prevent vignetting, the number of resolvable spots is reduced by the factor in brackets.

The relationship between $\Delta\phi_m$ (the maximum value of $\Delta\phi$) and L_1, which through (3.28) sets the maximum value of N_R, has been investigated by Lee and Zook [12]. From Fig. 3.6, ray optics gives

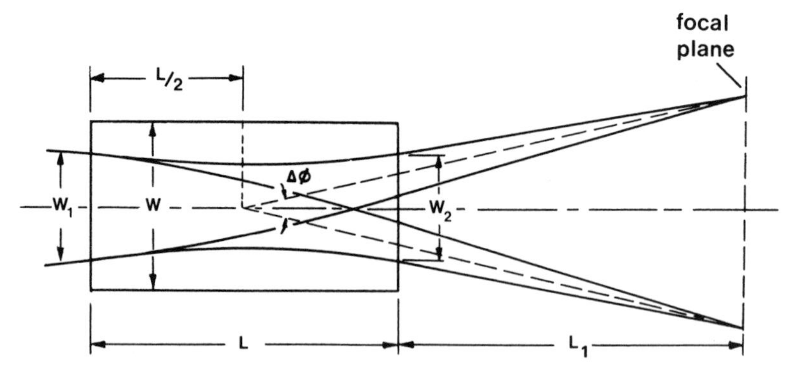

Figure 3.6 Prefocusing to avoid vignetting in an E-O deflector. (From Ref. 12.)

$$W_2 = \frac{\Delta\phi_m L}{2} + \frac{W_1 L_1}{L + L_1} \tag{3.29}$$

By eliminating L_1 from (3.28) and (3.29) and using $\theta_R = \lambda\varepsilon/nW_1$, they obtained

$$N_R = \frac{n\Delta\phi_m}{2\varepsilon\lambda}\left(W_1 + W_2 - \frac{\Delta\phi_m L}{2}\right) \tag{3.30}$$

This expression shows that N_R is a maximum when W_1 and W_2 are a maximum, that is, when $W_1 = W_2 = W$. Therefore,

$$N_R(\max) = \frac{\Delta\phi_m}{\theta_R}\left(1 - \frac{\Delta\phi_m L}{4W}\right) \tag{3.31}$$

Additionally, (3.29) can now be rearranged to give the beam focusing required to achieve $N_R(\max)$; that is,

$$\frac{L + L_1}{W} = \frac{2}{\Delta\phi_m} \tag{3.32}$$

Generally, only very weak focusing is required. As an example, if the deflector is capable of achieving $\Delta\phi_m \simeq 50$ mrad, Eq. (3.32) gives the focusing required as $\sim F/40$. For a typical deflector with an aspect ratio (L/W) of say 10, Eq. (3.31) shows that this focusing of the beam reduces N_R by only $\sim 12.5\%$ from that which could be obtained by using a collimated beam of width W in a deflector not limited in aperture.

Equations (3.31) and (3.32) optimize the deflector performance for a given maximum deflection $\Delta\phi_m$. If this angle is not limited by the maximum field that can be applied to the deflector, it is limited by the deflector aperture under the condition $L_1 = 0$; that is, when the beam is focused on the deflector exit aperture. Although under this geometrical limit the beam can still be considered quasi-parallel in each prism element, the beam diameter reduces such that in the last prism, where the beam is focused, the width is near zero. Hence, the induced optical path difference for the marginal rays tends toward zero in the prism elements near the deflector exit face. In terms of this limit, Lee and Zook [12] showed that N_R was given by

$$N_R = \frac{\Delta\phi_m}{\theta_R}\left(1 - \frac{1}{2}\beta^2\right) \tag{3.33}$$

Here $\beta = (L/W)/(L/W)_G$, where $(L/W)_G$ is the geometrically limited apect ratio for the deflection $\Delta\phi_m$.

To maximize N_R in (3.33) it was found necessary to fix some dimensional parameter for the deflector, for example, the aperture (W), area (LW), or volume. For these three specific cases, Lee and Zook [12] showed that N_R was maximized when β equaled $\sqrt{2/3}$, $\sqrt{2/5}$, and $\sqrt{1/3}$, respectively. In the last case, where N_R is maximized for a given volume, the deflector cost for a given performance is minimized, and the energy required to deflect the beam in a given time (which is proportional to volume but independent of shape) is also minimized.

In a deflector where $\Delta\phi_m$ is field rather than geometry limited, multiple passing can be used to increase N_R by increasing the deflector angle up to the optimum (β) limit. In a practical device, other factors, such as optical aberrations, reflection losses, or extractability of the deflected beam, will generally limit the number of passes that can be made.

A further parameter that affects N_R, and over which some control is possible, is the beam spatial intensity distribution. Equation (3.24) gives the diffraction-limited resolution angle θ_R in terms of the factor ε. For the case of a beam of uniformly intense spatial distribution used with a square aperture, $\varepsilon = 1.0$. If the aperture is circular, $\varepsilon \simeq 1.22$. Uniform illumination maximizes N_R, but in the far field diffracted energy produces cross-talk between adjacent beam positions. For the above apertures, this cross-talk is ~ 4.7 and $\sim 1.8\%$, respectively [15]. The effect is less severe if a truncated Gaussian beam is used. In particular, for a beam apertured at the $1/e^2$ intensity points, the cross-talk figures reduce to ~ 0.8 and $\sim 0.5\%$ for the two cases, respectively [12,16]. The penalty that has to be paid is an increase in ε to ~ 1.25 and ~ 1.27 for the respective apertures and consequent proportional decreases in N_R.

The above discussion has considered a perfect optic system used in conjunction with a focused beam of diffraction-limited divergence. In the real case, aberrations result from the use of a convergent beam with a deflector, and Beiser [17] has examined the effect of these on the resolution. Using ray optics, Beiser derived an expression for the focal spot size (d_a) due to the aberrations alone in terms of the F/No. and deflector refractive index. With the deflector immersed in an index-matching medium, this reduces to

$$d_a = \frac{n\Delta\phi(L/2 + L_1)}{16F^2} \qquad (3.34)$$

Where the device is not in an index-matching medium, additional aberration occurs at the deflector boundaries, and for this case Beiser found that d_a was twice as large.

The beam aberration will have little effect on the resolution so long as d_a is considerably smaller than the diffraction-limited spot size. If we let this ratio be δ, then under conditions where (3.31) applies, we have

$$\delta = \frac{n^2 \Delta\phi_m W}{16F^2 \varepsilon\lambda} \tag{3.35}$$

For the particular example given earlier, where $\Delta\phi_m$ = 50 mrad, F/No. = 40, and using $n \simeq 1.5$, $\varepsilon \simeq 1$, and λ = 632.8 nm, we obtain

$$\delta = 0.07 \text{ W cm}^{-1} \tag{3.36}$$

We see from this result that use of a weakly focused beam with a deflector of 1 or 2 cm in aperture does not significantly degrade the resolution. Additional care is needed when working with smaller F/No.'s, as besides causing loss of resolution, all aberrations adversely affect the cross-talk level of the deflector.

Figures of merit have been derived for both spatial resolution (M_R) and temporal response (M_τ) for linear E-O prism deflectors by Thomas [18]. For devices of square cross-section he obtained

$$M_R = \frac{\varepsilon_r \lambda N_R z}{wV_{app}} \quad \text{and} \quad M_\tau = \frac{\varepsilon_r \lambda \varepsilon_0 z}{I\tau} \tag{3.37}$$

where V_{app} is the potential applied across the deflector (in the z direction) to produce N_R resolved deflection positions, w is the usable aperture (i.e., $w < z$), and I is the current required by the device to allow adjacent spots to be resolved in the time τ. Thomas reviewed available linear E-O materials for use in a deflector, and his results are presented in Table 3.1.

For quadratic E-O materials, orientated so that the effective E-O coefficient was $g_{11} - g_{12}$, he derived comparative figures of merit:

$$M_R = \varepsilon_0^2(\varepsilon_r - 1)^2 n_0^2(g_{11} - g_{12})E_z$$

and (3.38)

$$M_\tau = \varepsilon_0^2 \varepsilon_r n_0^3(g_{11} - g_{12})E_z$$

Here ε_r is the relative dielectric constant for the crystal material, and E_z is a field of the same strength as used in the linear E-O material case.

Table 3.1 Summary of Some Possible Materials for Linear Deflector Use

Material	E-O coefficient r (10^{-12} m/V)	Relative dielectric constant, ε_r or ε_1, ε_3	n_0	Figure of merit: M_R, 10^{-9} m/V > 1.08	Figure of merit: M_R, 10^{-12} m/V > 9.5	Comment
ADP	24.5	56, 15	1.5	0.083	1.5	
KDP	10.3	44, 21	1.5	0.035	0.79	
KD*P	26.4	58, 50	1.5	0.089	1.5	
$LiNbO_3$	30.8 (r_{33})	80, 30	2.2	0.328	4.1	
$LiTaO_3$	30.3 (r_{33})	42.8	2.14	0.297	6.9	
$BaTiO_3$, const. strain (clamped)	840 (r_{42})	1970, 11	2.4	11.6	5.9	Very difficult to make stable, good-quality pieces
AMO	327 (r_{52})	17	1.5	1.1	65	
KTN	14,000 (r_{42})	Very high, ~ 5500	2.3	170	31	Cannot now be reliably made
HIO_3	?	20, 11	2	?	?	Very little known

Source: Ref. 18.

Generally, quadratic (including Kerr effect) materials were shown to have lower figures of merit than the best linear E-O materials in Table 3.1.

Finally it should be noted that, as with digital devices, two-dimensional deflection can be obtained with analog deflectors [12,13,19]. In this case, the second-stage deflection element, producing orthogonal deflection to the first, needs to be of comparable aperture in both deflection directions. As a consequence, a considerably larger deflection voltage and drive power are needed than in the first stage. Although Lee and Zook suggest a technique involving the use of thin crystal elements with interspersed electrodes to overcome this problem in the second stage, digital deflectors are more attractive for many two-dimensional applications, particularly those requiring a large number of resolved beam positions. As a consequence, little work has been published in the literature on comparable two-dimensional analog devices.

3.2.2 Refractive Index Gradient Deflectors

Operating Principle

Whereas homogeneous refractive index changes only result in optical deflection at prism interfaces, gradiential changes also produce deflection in the bulk material. Two common examples of the latter effect are the bending of radio waves in the ionosphere and the refraction of the sun's rays in the earth's atmosphere.

Deflection by an E-O crystal, due to a refractive index gradient, was first reported by Fowler et al. [20]. They showed that a linear refractive index gradient, transverse to the direction of optical beam propagation, results in a deflection that is proportional to both the index gradient and interaction length. This can be seen with the help of Fig. 3.7.

Fowler et al. assumed that the refractive index changes induced in the medium were so small that the light rays traveled in nearly straight lines. For a beam incident normally on the end face of the deflector, rays A and B arrive in phase but travel through the medium with different velocities. If the refractive indices for the two marginal rays are $n_0 + \Delta n$ and $n_0 - \Delta n$, respectively, a phase difference $\Delta\psi$ between them of $4\pi \Delta n L/\lambda$ will have accumulated at the output face. As a result, the output beam wavefront becomes CD, where d in the figure is a distance equivalent to the phase lag $\Delta\psi$; that is,

$$\Delta\psi = \frac{2\pi n_2 d}{\lambda_0} = \frac{4\pi \Delta n L}{\lambda_0} \qquad (3.39)$$

Therefore,

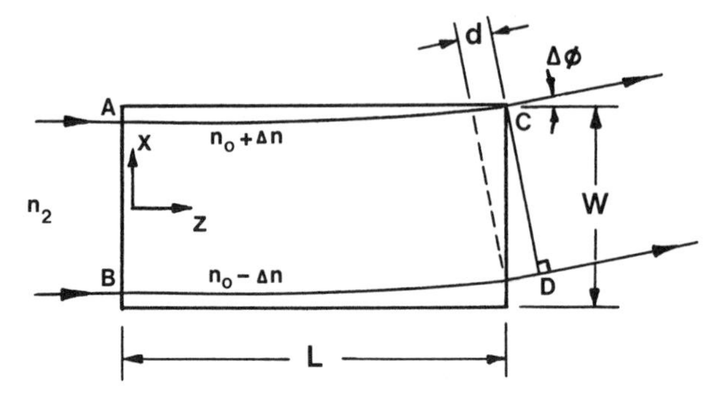

Figure 3.7 Passage of an optical beam through a gradient deflector. (From Ref. 20.)

$$d = \frac{2L\,\Delta n}{n_2} \tag{3.40}$$

The above analysis considers the *small-angle* limit, and consequently the deflection $\Delta\phi$ is given by

$$\Delta\phi \simeq \sin\Delta\phi \simeq \frac{2L\,\Delta n}{n_2 W} \tag{3.41}$$

By using (3.24), the number of resolved beam positions (N_R) is therefore

$$N_R = \frac{\Delta\phi}{\theta_R} = \frac{2L\,\Delta n}{\varepsilon\lambda} \tag{3.42}$$

It can be seen that the last two results are identical to those given by (3.26) and (3.27) for the deflection and number of resolved beam positions, respectively, of an E-O prism-type deflector.

A more rigorous analysis of beam propagation in a gradient deflector has been carried out by Grieb et al. [21] using the calculus of variations (see, for example, Born and Wolf [15] and Stavroudis [22]). This approach is a generalization of ordinary geometrical optics and is the mathematical basis for Fermat's principle. That is, in the coordinates used here, it essentially reduces the problem to finding the ray path for which the integral I given by

$$I = \int_0^L n\left(1 + \left|\frac{dx}{dz}\right|^2\right)^{1/2} dz \tag{3.43}$$

is an extremum (i.e., maximum or minimum).

Grieb et al. considered the two distinct cases where the E vector of the incident beam was either perpendicular to or in the plane of deflection (in the former case, the bending of the beam does not affect the polarization orientation, but in the latter case it does). Grieb and co-workers showed that, in the small-angle limit, both beam polarizations resulted in near parabolic ray trajectories, confirming the accuracy of the linear result given by (3.41). They deduced the ratio (δ) of the output to input beam divergence and obtained

$$\delta = \left[1 - \left(\frac{\Delta nL}{n_0 W}\right)^2\right] \tag{3.44}$$

For achievable refractive index changes (i.e., $\Delta n \lesssim 10^{-3}$) this ratio is within $\sim$ 0.01% of unity. Consequently, a gradient deflector has negligible effect on the beam divergence. This result is implicit in (3.41), where it can be seen that beam deflection is only a function of L. That is, in the small-angle limit, deflection does not depend on ray position in the wavefront.

Grieb et al. generalized their analysis of the gradient deflector and considered the case where the refractive index was some arbitrary function of the transverse coordinate x. They found that the resultant deflector was determined only by the relative change in refractive index over the beam width. Specifically, for the case where the refractive index varied according to x^q ($q > 1$), they calculated that the deflection dimensions required to obtain a given deflection became smaller than in the linear gradient ($q = 1$) case but that the beam aberrations were worse.

Electrode Profile

Electrodes in a quadrupole arrangement are generally employed with a gradiential deflector. To provide a linear refractive index gradient in a linear E-O material, they need to be of hyperbolic profile [14,23]. Electrodes of this section are difficult to fabricate with the accuracy required. Consequently, calculations have been carried out to determine the effect on field linearity of replacing them by simply cylindrical electrodes. Figure 3.8 shows some computed curves of the field error ($\Delta E/E$) resulting from the use of cylindrical electrodes (radius R) to match a hyperbolic profile [23]. Although the results presented in the figure relate specifically to the hyperbola $xz = 1$ (where x and z are in millimeters), the dimensions can be scaled linearly to fit a deflector of arbitrary aperture. The curves show that use of cylindrical electrodes with a gradient deflector can result in a substantial (> 10%) field error. As a result of this error the wavefront of the optical beam is distorted as it propagates through the deflector. For a beam deflected through N_R spots, the accumulated wavefront error ($\Delta\lambda$)

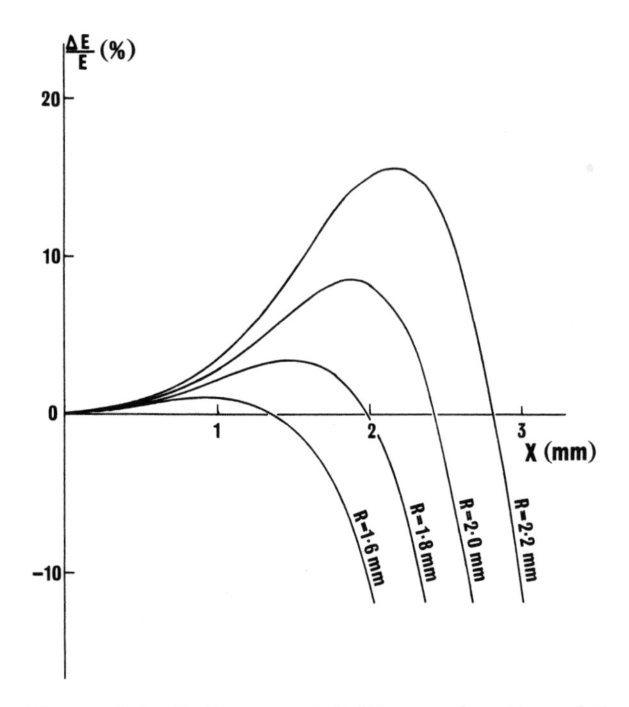

Figure 3.8 Field error (ΔE/E) as a function of X arising from the use of electrodes of cylindrical section (radius R) rather than of hyperbolic profile to form a quadrupole field for a crystal deflector. (From Ref. 23.)

after transit through the crystal is $(2N_R \varepsilon x/w)(\Delta E/E)$ wavelengths. Figure 3.9 is a plot of this error for the specific case discussed above with R = 2.0 mm, w = 3.0 mm, and N_R taken as 40 [23]. Although the field distortion leads to a wavefront error of several wavelengths, the figure shows that it can be represented by a linear component (that simply adds extra deflection to the beam) and a much smaller component, which in this example is only of order $\lambda/2$.

Similar conclusions concerning the relative merits of cylindrical and hyperbolic electrodes for a gradient deflector have been drawn by Sevruk and Gusak [24] but for a different reason. They examined in detail the effect of finite deflector dimensions on the field linearity. The calculations, based on the numerical solution of the Laplace equation, showed that replacement of hyperbolic electrodes by ones of circular section did not necessarily lead to appreciable loss of deflector performance. This conclusion is a result of edge effects in a deflector of limited transverse dimensions distorting the (ideal) field of hyperbolic electrodes to such an extent as to render them no advantage over

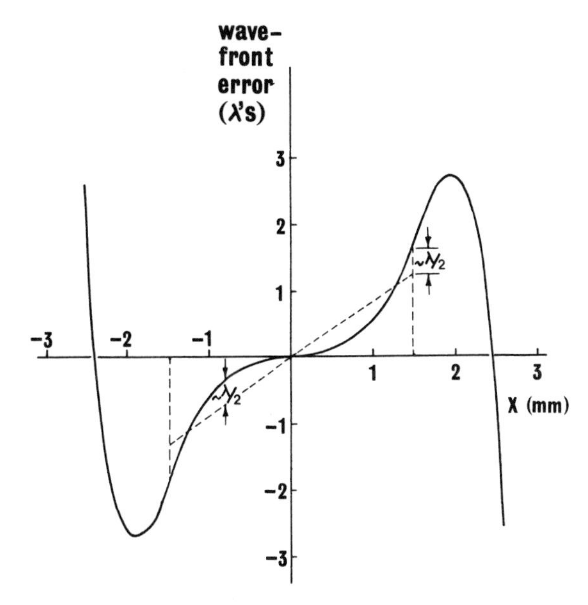

Figure 3.9 Wavefront error resulting from single transit of a beam through a gradient deflector with cylindrical electrodes. In this case, R = 2.0 mm, w = 3.0 mm, and N_R has been taken as 40. The error results in a small additional beam deflection plus a ~ λ/2 wavefront aberration. (From Ref. 23.)

ones of optimum circular section. For the particular case of a deflector in air and with transverse dimensions limited to 1.5 times the clear aperture, they computed the radius of circular section electrodes that minimized the average field error over the circular beam aperture. For three commonly used crystal orientations, Sevruk and Gusak showed that an appropriate choice of electrode radius could reduce the average field error to only a few percent.

Frequency Response

In a deflector used for scanning or streak applications the refractive index is a function of both transverse coordinate and time; that is, n = n(xt). The calculus of variations can be used to show that for streak applications the linear approximation for the deflection is still valid. This is not intuitively the case, for at high sweep speeds the expected time resolution (time to sweep through one resolvable spot) can be less than the transit time of the beam through the deflector, and this could possibly be significant.

For a deflector with a linear refractive index gradient which is also a linear function of time, the refractive index can be written as

$$n(xt) = n_0(1 + \alpha xt) \tag{3.45}$$

where n_0 and α are constants. If $t = t_0 + t_z$, where t_0 is the time at which a ray reaches the input face of the deflector after the start of the linear field ramp and t_z is the time taken by the ray to reach the point (xz) in the crystal, then in the small-angle limit, $t_z = zn_0/c$. Consequently, (3.45) becomes

$$n(xz) = n_0\left[1 + \alpha x\left(t_0 + \frac{n_0 z}{c}\right)\right] \tag{3.46}$$

Here the refractive index is only a function of spatial coordinates. The integral I given by (3.43) can be formed and the integrand used to set up the Euler equation from which the external space curve, representing the ray trajectory in the deflector, can be found. The gradient at the output face is equal to $(n_2/n_0)\Delta\phi$. Consequently, differentiation with respect to t_0 at $z = L$ gives the linearity of the deflector response. This procedure yields

$$\frac{d(\Delta\phi)}{dt_0} = \frac{n_0 \alpha L}{n_2}\left(1 - \frac{1}{2}\alpha^2 L^2 t_0^2\right) \tag{3.47}$$

To assess its significance, the last term in (3.47) can be determined for a typical deflector. For a ~ 5-cm-long device of a few millimeters in aperture achieving a refractive index change of 10^{-3} in say 5 nsec, the term is < 0.01. This result shows that a gradient deflector can be operated at high streak speeds corresponding to a few picoseconds of resolution without the optical transit time leading to nonlinear performance.

In the case of an oscillatory applied field, as in a device used for scanning, the deflector cannot respond in times which are comparable to, or less than, the optical transit time through the crystal. This limit arises since the final deflection is the result of an *integrating* effect as the optic beam passes through the crystal.

For the example of the ~ 5-cm-long deflector discussed above and made (say) from $LiNbO_3$, the maximum scanning frequency would be restricted to ~ 500 MHz. Correspondingly, in a streak application, the linear field ramp must be applied for a period long compared with that of the optic transit time to achieve the linear performance limit of (3.47).

In the case of a prism deflector a similar constraint on scanning frequency applies. The equivalent streak case has been examined in detail by Elliot and Shaw [25]. The fact that the deflection is localized at the interface of the two prisms means that different parts of the wavefront of the propagating beam are refracted at different times after they enter the deflector. As a result, application of a voltage ramp produces the greatest deflection in the parts of the wavefront reaching the interface latest. This leads to wavefront curvature. In scanning applications, the same frequency limit as that given above for the gradient deflector applies, since otherwise the wavefront curvature becomes excessively aberrated.

As a result of the finite transit time, attempts to operate a gradient deflector well above the maximum frequency limit would result in a static, undisplaced spot. For a simple prism deflector the result would be an extended line image over the total scan range.

The field transit time is a further important consideration with implications for the high-frequency performance of an analog deflector. Table 3.1 shows that relative dielectric constants of $\sim$ 50 are typical of linear E-O crystals. Consequently, the field propagation velocity is usually considerably slower than that of the optic beam. Techniques, such as matching the deflector into a broad-band transmission line, can help distribute the current more uniformly at high frequencies. For some E-O materials it is possible to achieve collinear and synchronous propagation of the optic and electric fields [25-27]. This makes a device which is potentially very broad band.

Optimized Design

The comments made earlier concerning optimization of the design of a prism-type analog deflector are generally applicable to a gradient device. One difference is the figure of merit for temporal response given in (3.37). In a gradient deflector, the capacitance depends on the particular electrode geometry used and is usually greater than in the equivalent iterated prism device. In a quadrupole arrangement, the electrodes generate a field gradient in the direction orthogonal to the deflection field, and consequently the relative dielectric constant of the E-O material in this direction is equally important when assessing the temporal response of the device and drive power requirement.

3.2.3 Analog-Digital Array Deflectors

In Sec. 3.2.1 it was shown how a series arrangement of analog deflector elements can be used to increase N_R by increasing $\Delta\phi$. From the definition of N_R in Sec. 3.2.1, it can be seen that an alternative means of increasing N_R is to reduce the beam diffraction angle θ_R. As the minimum value of θ_R is determined by the beam width, this approach implies an increase in deflector aperture.

Figure 3.10 is an example showing how a parallel array of prism-type deflector elements can be used to make a large-aperture device [28]. In this case, each deflector element produces a deflection of $\Delta\phi_1$ given by Eq. (3.26):

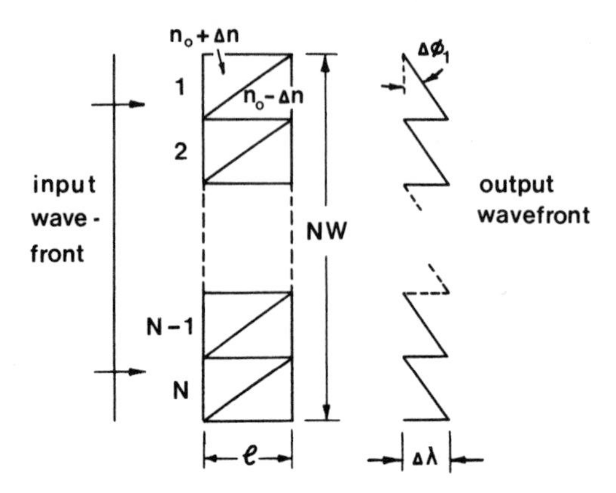

Figure 3.10 Analog-digital array deflector using prism-type deflector elements. (From Ref. 28.)

$$\Delta\phi_1 = \frac{2\,\Delta n \ell}{n_2 W} \tag{3.48}$$

Since there are N parallel elements, the minimum diffraction angle θ_R is given by

$$\theta_R = \frac{\varepsilon\lambda}{NWn_2} \tag{3.49}$$

Consequently, the maximum number of resolved beam positions that it is possible to achieve is

$$N_R = \frac{\Delta\phi_1}{\theta_R} = \frac{2\,\Delta n \ell}{\varepsilon\lambda} N \tag{3.50}$$

that is, N times that of a single-prism element.

It can be seen from the figure that the parallel array deflector behaves as a phase grating, and in general the phase of the transmitted beam is discontinuous. Ninomiya [28] calculated the Fraunhofer (far-field) diffraction pattern expected from this deflector and came to the following conclusions:

1. Without bias voltages on the deflector elements, the deflection $\Delta\phi$ is discontinuous.
2. When $\Delta\lambda = M\lambda$ (M equal to a positive integer), the beam is deflected perfectly in the direction $\Delta\phi$, given by

$$\Delta\phi = \frac{M\lambda}{Wn_2} \tag{3.51}$$

3. Continuous deflection requires a bias voltage, increasing in increments, to each element in the array such that for the mth element it produces a retardation $\Delta\lambda_m$ given by

$$\Delta\lambda_m = (m-1)\,\Delta\lambda - q\lambda \tag{3.52}$$

where q is zero or an integer.

4. Discrete deflection angles $\Delta\phi_t$, given by

$$\Delta\phi_t = \frac{t\lambda}{NWn_2} \tag{3.53}$$

where t is an integer, result if the bias voltage produces a retardation $\Delta\lambda_m$, where

$$\Delta\lambda_m = \frac{t}{N}(m-1)\lambda - q\lambda \tag{3.54}$$

In the digital case (4), the angle between adjacent deflection positions is λ/NWn_2 and can be considered a *digital unit*. However, since the smallest resolved angle is θ_R, given by (3.49) with $\varepsilon > 1$, an angle of two-digit units is resolvable, but generally one is not.

Parallel array deflectors are potentially attractive in applications requiring either high deflection sensitivity or a large number of deflection positions, either analog or digital. In particular, the planar geometry makes these devices suitable for use in integrated optic circuits. In this application, diffusion techniques can be used to make a shallow waveguide layer of depth ~ 100 μm in E-O materials and surface electrodes used to provide the deflection field [29,30] (see Chap. 4).

In a recent review, Bulmer et al. [31] have examined some possible geometries for waveguide array deflectors. They found that the far-field intensity distributions are given by the product of two terms. One is an array function (AF), which is determined by the period of the array and the number of elements it contains and is independent of the phase distribution within any one element. The other is the element function (EF), which is the square of the magnitude of the Fourier transform of the transmission function at the output aperture of one element. Whether the device produces a deflection that is continuous or not depends on whether the EF or AF, or both, is displaced as a function of voltage. If a phase slope is created across the wave propagation through one element, as for the deflector in Fig. 3.10, the resulting element pattern is deflected as a function of the voltage inducing the phase slope. If there is a phase difference between consecutive elements in an array, the array pattern is deflected as a function of that phase difference. This is the case for the prism array when bias voltages are applied to the individual deflector elements. Bulmer and co-workers tabulated the type of deflection expected from the different geometries that they investigated.

For a prism array, Revelli [32] has shown that the geometric limit to the number of resolvable spots $N_R(\max)$ is set by total internal reflection at the output face. He obtained

$$N_R(\max) = \frac{2}{\theta_R}\left(\frac{F-1}{F+1}\right)^{1/2} \tag{3.55}$$

where $F = n(n^2 - 1)^{1/2}$ and n is the refractive index of the crystal. When N_R is large, the small-angle minimum deviation approximation of Sec. 3.2.1 is no longer valid and (3.24) must be used for θ_R. This makes the effective aperture of the device a function of the deflection angle. As an example of the limit represented by (3.55), Revelli considered an $LiNbO_3$ deflector with aperture NW = 1 cm. For $\Delta n/n = 10^{-3}$, he found $N_R(\max) \simeq 10^3$.

3.3 OTHER E-O DEFLECTION TECHNIQUES

Although the digital and analog light deflectors discussed in the previous sections have received the most attention for high-frequency deflection and scanning applications, several other E-O deflector schemes have been demonstrated or proposed in recent years. The following are some of the more interesting.

3.3.1 Analog Deflector Using Frequency Shifting

A novel deflector involving frequency shifting has been proposed and demonstrated by Wilkerson and Casperson [33]. In this device an E-O crystal is used to impress a frequency shift on an optical beam which is subsequently deflected by a dispersive element, that is, a prism, diffraction grating, or Fabry-Perot etalon. Although similar to a DLD in that it is a two-element device with the switching and deflection functions separated, the deflection in this case can be continuous since arbitrary frequency shifts are possible.

In a linear E-O material a frequency shift is obtained by applying a linear voltage ramp. The resultant refractive index sweep causes the phase of an optical pulse propagating through the material to vary linearly with time. Since a linear phase variation is equivalent to a frequency shift, the pulse leaving the crystal is centered at a different frequency from that when it entered. The dispersive element following the frequency shifter deflects the beam by an amount which depends on the frequency shift achieved.

For a frequency shifter with a weak time-dependent index of refraction of the form

$$n = n_0 - \Delta n \sin \omega t \tag{3.56}$$

where w is the angular frequency of the modulator, it can be easily shown that for a portion of the light pulse crossing the center of the crystal at time t the instantaneous frequency shift is

$$\Delta \nu(t) = \frac{\omega \nu d \, \Delta n}{c} \frac{\sin(2\omega T)}{2\omega T} \cos \omega t \tag{3.57}$$

where ν is the central frequency of the optical beam, d is the length of the crystal, and T is the transit time. If the light pulse is short compared to the RF modulation period and is synchronized to cross the crystal when ωt is an integral multiple of 2π, then $\cos \omega t$ may be approximated by unity, and the entire pulse is upshifted by the maximum amount. The factor $\sin(2\omega T)/2\omega T$ is also equal to unity if the optic transit time is short compared to the modulation period or if appropriately matched traveling-wave modulation fields are employed. Consequently, the maximum frequency shift is

$$\Delta \nu \simeq \frac{\omega \nu d \, \Delta n}{c} \tag{3.58}$$

If the dispersive element is chosen to just resolve one full bandwidth of the optical pulse spectrum, then the number of resolvable spots for the beam scan is just the number of spectral bandwidths that can be fitted into the total frequency shift given by (3.58). For an optic pulse of duration t_p that is transform-limited, that is, $\nu_p t_p \simeq 1$, the number of resolvable spots (N_R) becomes

$$N_R = \frac{\Delta \nu}{\nu_p} \simeq \Delta \nu t p \simeq \frac{\omega \nu d t_p \, \Delta n}{c} \tag{3.59}$$

For the example of 1-nsec duration bandwidth-limited pulses from an He-Ne laser operating at 633 nm being frequency-shifted in a ~ 5-cm-long crystal modulated at 50 MHz and achieving a maximum refractive index change of 10^{-3}, (3.59) gives $N_R \simeq 25$. In fact, since both $\pm\Delta n$ refractive index changes are possible, the direction of frequency shift can be reversed, and the maximum number of resolved beam positions becomes $2N_R$. It can be seen from (3.59) that further increase in the number of resolved positions is possible by maximizing the crystal length, modulating at a higher frequency, or employing multipass techniques, provided, of course, that the approximations resulting in (3.58) remain valid. The maximum theoretical access time of the deflector is the crystal transit time and would be achieved with a *sawtooth* applied field of this period.

3.3.2 Combined E-O and Acoustic Digital Deflector

In Sec. 3.2 the operating principles of prism deflectors were discussed. It was shown that in bulk single-element devices the achievable refractive index change produced only limited deflection capability. Nevertheless, in the form of miniature surface-wave devices these simple deflectors have important potential applications, because the scaling down of their size for inclusion in integrated optic circuits results in the required deflection field being achieved by only a modest applied voltage (see Sec. 4.3.2). For these integrated circuit applications there is a continuing search for techniques to minimize the required drive voltage to allow operation of the deflector by simple transistor logic. The localized nature of the power dissipation in the drive transistors and deflector elements can also present problems for high-frequency operation if the voltage is not minimized.

A simple scheme to reduce the deflection drive voltage has been proposed by Kotani et al. [34]. These workers proposed the combination of an E-O prism deflector with an acousto-optic grating deflector and showed that potentially an order-or-magnitude reduction in drive voltage for switching between two beam directions was achievable. A schematic of the proposed deflector arrangement is shown in Fig. 3.11.

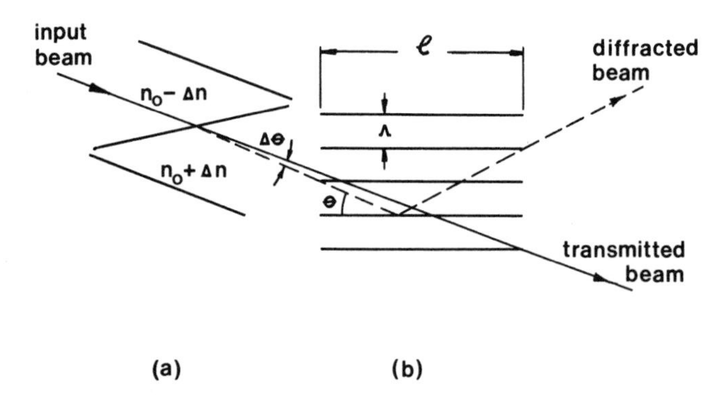

Figure 3.11 Combined E-O and acoustic digital deflector. (a) E-O from deflector, (b) acoustic deflector operating in the Bragg regime.

A surface corrugation grating diffracting in the Bragg regime couples the maximum energy into the diffracted beam when the *Bragg condition* is satisfied, that is, when

$$\Lambda \sin \theta = \frac{\lambda}{2n} \qquad (3.60)$$

Here Λ is the grating period, θ the incident ray angle, λ the free-space wavelength, and n the effective index of the guided optical mode.† From (3.60) it can be seen that the coupling of energy into the diffracted beam can be controlled by either a change in n or θ. Kotani et al. [34] showed that the diffracted energy was much more sensitive to changes in θ than in n. For example, they calculated that for a grating with spacing Λ of 1 μm and length ℓ of 5 mm the required angular change $\Delta\theta$ for complete switching was 170 μrad. In principle, this deflection can be achieved in a single-element surface-wave $LiNbO_3$ prism deflector with an applied voltage of only $\sim$ 1.5 V. In comparison, the same deflector would require 14 V if it operated alone to produce two well-resolved beam positions.

To achieve the maximum performance from the combined prism and grating deflector, Kotani and co-workers calculated that the grating spacing would need to be very tightly controlled and the grating refractive index very homogeneous. For the above example, they calculated that the grating spacing error $\Delta\Lambda$ should be << 12 Å and that the refractive index inhomogeneity Δn should be $<< 2.6 \times 10^{-3}$.

To avoid cross-talk, the diffraction beam spread must be less than the change in angle. Since $\sim$ 170 μm corresponds to the divergence from a diffraction-limited Gaussian beam of $\sim$ 100-μm diameter at 633 nm, this sets the limit to the miniaturization of a device based on this principle.

3.3.3 Analog Interference Deflectors

In principle, the deflectors discussed in Secs. 3.1 and 3.2 could be used with any approximately monochromatic light source so long as it was spatially filtered to produce a beam of divergence close to the diffraction limit. In practice, beams from lasers are generally used since only they can provide the intensity required in most deflection and scanning applications. In common with the analog deflector discussed in Sec. 3.3.1, the interference deflector makes use of the other principal laser property, high monochromaticity. In principle, the use of multiple-beam interference effects allows scanning devices to be built which require a change in optical path of only half a wavelength per

---

†See Chap. 10 for a more complete discussion of these devices and the derivation of the equations that characterize their operation.

transit to scan through all the resolved positions. This contrasts with other types of E-O deflectors, which for a similar change in refractive index in the active material produce only a deflection of one resolved position.

The interference deflector is based on the Fabry-Perot interferometer and is shown schematically in Fig. 3.12. The far-field beam pattern from the interferometer is a series of narrow fringes whose intensity distribution I(x) is proportional to the following function:

$$I(x) \propto \left(1 + F \sin \frac{\psi}{2}\right)^{-1} \tag{3.61}$$

where the phase difference between consecutive transmitted beams $\psi = 4\pi/\lambda(d + \beta x)n \cos \alpha$ and F is the finesse of the interferometer. It can be seen from (3.61) that the spacing between adjacent fringes in the output distribution depends on the wedge angle β. This angle

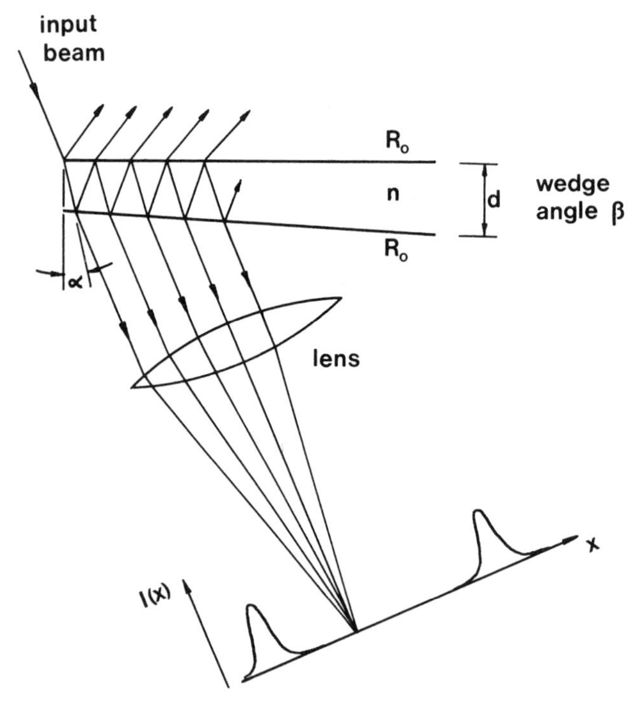

Figure 3.12 Fringes formed by beams multiply reflected and transmitted by an interferometer comprising two high-reflectivity mirrors at a small wedge angle. The etalon is the basis of an interference deflector.

can be chosen so that only one fringe appears across the desired linear field at the lens focal plane. A change in phase corresponding to one-half wavelength in the cavity (i.e., ψ changes by 2π) results in the fringe being scanned completely across the linear field range. This phase change can be produced either by varying the etalon spacing d or by changing the refractive index n. The number of resolvable elements N_R in the scan is determined by the ratio of the fringe width to the fringe spacing. This ratio can be either aperture on loss limited [35]. For a sufficiently monochromatic beam Korpel [35] has shown that the maximum value of N_R is equal to the etalon finesse:

$$N_R = \frac{\pi\sqrt{R_0}}{1 - R_0} \tag{3.62}$$

where R_0 is the reflectivity of each of the mirrors. For mirrors with $R_0 \sim 99.7\%$ we see that N_R can exceed 10^3.

Buck and Holland [36] have noted that the scanning rates achievable by these devices are ultimately limited by the time needed to reestablish the field in the etalon after deflection to a new position. This time increases with the etalon finesse and consequently with the resolution of the deflector. In principle, the etalon spacing (d) can be made arbitrarily small to minimize this time.

3.3.4 Analog Deflector Using Frequency Tuning

While the scanning technique in 3.3.1 relies on a narrow bandwidth (ideally transform limited) laser for the attainment of high resolution, a similar deflection scheme, but appropriate for use with wide bandwidth lasers (e.g., dye or Alexandrite lasers) has recently been proposed by Filinski and Skettrup [37]. In this case it has been proposed that the frequency shifting be achieved by the use of an intracavity E-O tuning element [38] rather than a frequency modulator. The result is a narrow bandwidth optical output that is tunable over the broad laser bandwidth. As in Wilkerson and Casperson's scheme, scanning is achieved by following the laser with a high dispersion element.

For the specific case of an Alexandrite laser with an assumed useful tuning range of 120 nm, Filinski and Skettrup calculated that a deflection of > 3000 resolvable spots could be achieved by use of a suitable dispersive grating and that a deflection rate of up to $\sim 10^{10}$ spots per second was possible. The rate attainable being ultimately limited by the time taken for the laser to re-establish a steady-state E-M field distribution and suitably narrow bandwidth at the new frequency.

REFERENCES

1. T. J. Nelson, *Bell Syst. Tech. J. 43*:821 (1964).
2. W. Kulche, K. Kosanke, E. Max, M. A. Haerger, T. J. Harris, and H. Fleischer, *Appl. Opt.* 5:1657 (1966).
3. J. C. Bass, *Radio Electron. Eng.* 345 (Dec. 1967).
4. S. Flugge, *Handb. Phys. 24*:431 (1956).
5. R. Pepperl, *Opt. Acta 24*:413 (1977).
6. W. Kulcke, K. Kosanke, E. Max, H. Fleisher, and J. J. Harris, *Optical and Electro-Optical Information Processing.* M.I.T. Press, Cambridge, Mass.: (1965), Chap. 33.
7. J. M. Ley, T. M. Christmas, and C. G. Wildey, *Proc. IEEE 117*:1057 (1970).
8. U. Kruger, R. Pepperl, and U. Schmidt, *Proc IEEE 61*:992 (1973).
9. U. J. Schmidt, E. Schroder, and W. Thust, *Appl. Opt. 12*:460 (1973).
10. S. K. Kurtz, *Bell Syst. Tech. J. 45*:1209 (1966).
11. I. P. Kaminow, *Appl. Opt.* 3:511 (1964).
12. T. C. Lee and J. D. Zook, *IEEE J. Quantum Electron. QE-4*:442 (1968).
13. F. S. Chen, J. E. Geusic, S. K. Kurtz, J. G. Skinner, and S. H. Wemple, *J. Appl. Phys. 37*:388 (1966).
14. J. F. Lotspeich, *IEEE Spectrum,* 45 (Feb. 1968).
15. M. Born and E. Wolf, *Principles of Optics,* 5th ed., Pergamon, London (1975).
16. A. L. Buck, *Proc. IEEE (Lett.),* 448 (March 1967).
17. L. Beiser, *J. Opt. Soc. Am. 57*:923 (1967).
18. S. W. Thomas, *Proc. 13th Int. Congr. on High Speed Photogr. and Photonics, Tokyo, 1978,* Vol. 189, SPIE (1978), p. 499.
19. J. D. Beasley, *Appl. Opt. 10*:1934 (1971).
20. V. J. Fowler, C. F. Buhrer, and L. R. Bloom, *Proc. IEEE (Corres.),* 1964 (Feb. 1964).
21. B. N. Grieb, P. A. Korotkov, and V. N. Mal'nev, *Sov. Phys. J. USA 12*:1207 (1976).
22. O. N. Stravroudis, *The Optics of Rays, Wavefronts and Caustics,* Vol. 38 in series on Pure and Applied Physics, Academic, New York (1972).
23. C. L. M. Ireland, *Opt. Commun. 30*:99 (1979).
24. B. B. Sevruk and N. A. Gusak, *Opt. Spektrosk 45*:910 (1978).
25. R. A. Elliot and J. B. Shaw, *Appl. Opt. 18*:1025 (1979).
26. I. P. Kaminow and E. H. Turner, *Proc. IEEE 54*:1374 (1966).
27. A. A. Basov, A. A. Vorob'yev, and I. G. Katayev, *Radio Eng. Electron. Phys. 22*:77 (1977).
28. Y. Ninomiya, *IEEE J. Quantum Electron. QE-9*:791 (1973).
29. I. P. Kaminow and L. W. Stulz, *IEEE J. Quantum. Electron. QE-11*:633 (1975).

30. C. S. Tsai and P. Saunier, *Appl. Phys. Lett.* *27*:248 (1975).
31. C. H. Bulmer, W. K. Burns, and T. G. Giallorenzi, *Appl. Opt.* *18*:3282 (1979).
32. J. F. Revelli, *Appl. Opt.* *19*:389 (1980).
33. J. L. Wilkerson, and L. N. Casperson, *Opt. Commun.* *13*:117 (1975).
34. H. Kotani, S. Nambo, and M. Kawabe, *IEEE J. Quantum Electron.* *QE-15*:270 (1979).
35. A. Korpel, *Proc. IEEE (Corres.)* *53*:1666 (1965).
36. W. E. Buck and T. E. Holland, *Appl. Phys. Lett.* *8*:198 (1966).
37. I. Filinski, and T. Skettrup, *IEEE J. Quantum Electron.* *QE-18*:1059 (1982).
38. J. M. Telle and C. L. Tang, *Phys. Lett.* *24*:85 (1974).

4

Electro-Optic Deflector Designs

4.1 DIGITAL LIGHT DEFLECTORS (DLDs)

4.1.1 General

The DLD uses a birefringent element to give spatial separation between switched orthogonal polarization states of an incident light beam. The birefringent elements used in DLDs have been calcite rhombs or prisms and the polarization switches electro-optic cells of the liquid Kerr or the solid-state Pockels type. For this important deflector there is only a limited choice of components. Proposed or used birefringent elements are shown in Fig. 3.1 and polarization switches in Fig. 4.1.

A purely theoretical approach to the selection of optic components for the DLD is unlikely to achieve a practical device, because the optic components may not be readily available and the problems associated with their use may not be well documented or necessarily well understood. DLDs for which published information exists have all required the practical development of the chosen optic components. Consequently, work in this field has required access to many practical skills and in particular those necessary for the production of crystal and liquid optics.

4.1.2 Kerr Cell Deflectors

Probably the most advanced and successful work on DLDs has been carried out at Philips GmbH Forschungslaboratorium Hamburg by Schmidt [1] and others as part of a program to develop a large-screen laser display facility. This work started in 1964, and by 1972 a 20-stage light beam deflector giving a two-dimensional raster of 1024 × 1024 positions had been successfully operated [2]. This system has not been outdated in performance by any later design and is therefore still of considerable interest. Schmidt's 20-stage deflector is also of special interest in that it is an integrated design using only nitrobenzene and calcite as the active optic components. There are no air interfaces. The deflector is divided into two sections of 10 stages and is shown in Fig. 4.2.

The deflector built by Schmidt embodied the following design features.

The refractive index of nitrobenzene, which is very close to the mean of the refractive indices of calcite, removed the requirement for

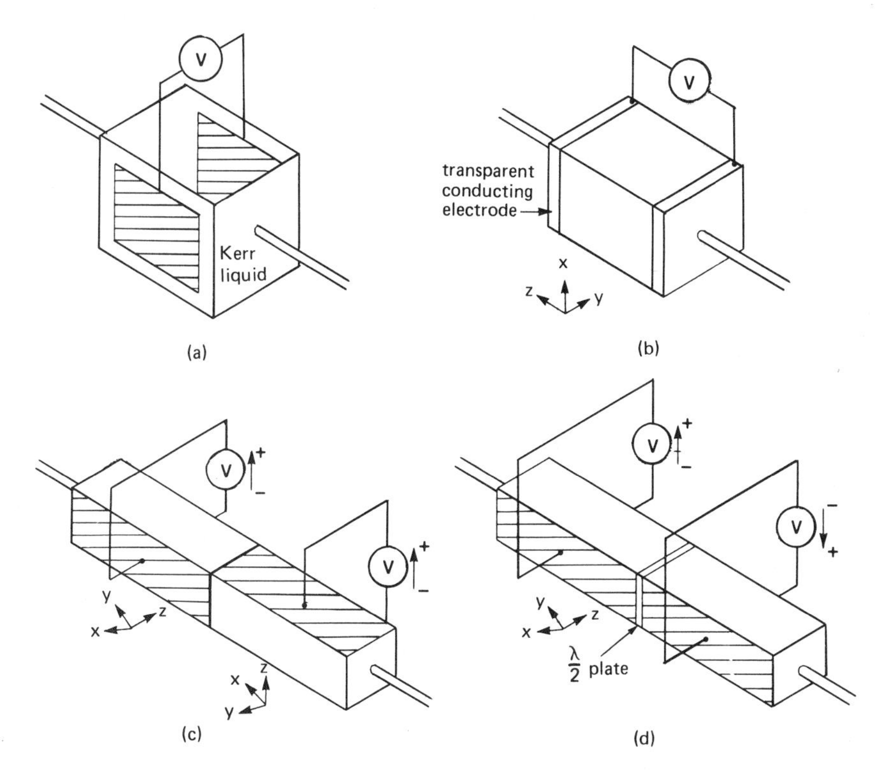

Figure 4.1 Electro-optic polarization switches. (a) Kerr cell nitrobenzene [8]; (b) longitudinal Pockels cell; KD*P Z-cut with conducting transparent electrodes; (c) transverse composite without half-wave plate; KD*P 45° Z-cut [4]; (d) transverse composite with half-wave plate; KD*P 45° Z-cut.

compensating wedges to balance the ordinary and extraordinary ray paths about the systems axis.

The first 10 stages of deflection achieved a 32 × 32 block scan using calcite prisms of 6' 12", 24' 48", and 1°36'. (Refer to Table 4.1.) This design was achieved with an electrode separation of only 1.4 to 1.5 mm and consequently required a relatively low switching voltage of 2.1 kV. The deflector gave a high-speed localized writing facility of particular value in alphanumeric and graphic displays.

The second deflection stage of 10 units was coupled to the first by a lens system designed to achieve the most favorable electrode spacing at the final stages of deflection. In this way the drive voltage of the last two stages was kept below 10 kV for an electrode spacing of 6 mm. A degree of compensation for beam wander, due to the variation of refractive indices of calcite and particularly nitrobenzene with tempera-

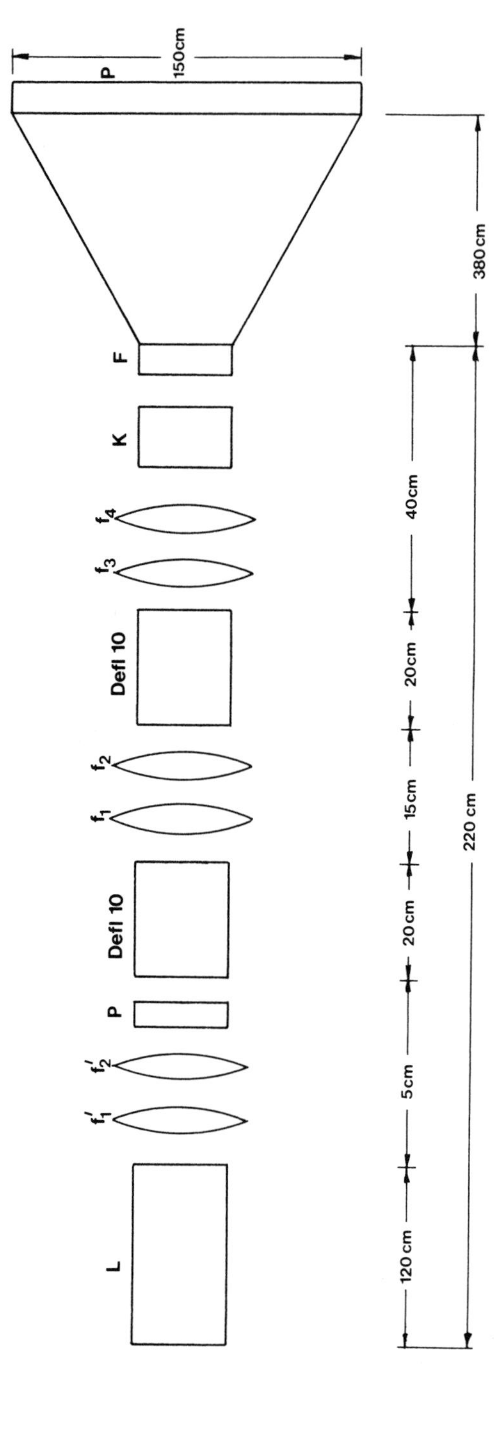

Figure 4.2 Optic section of the 20-stage deflector used in an experimental large-screen projector for alphanumeric data. L, laser; P, polarizer; Defl 10, deflector with 10 stages; f_1' (40 mm) and f_2' (80 mm), telescope pair; f_3 (90 mm) and f_4 (35 mm), projection optics; K, Kerr cell; F, polarization filter; f_1 and f_2, matching lenses; P, projection screen. (From Ref. 1.)

Table 4.1 Design Data for a 20-Stage Deflector

Stage no.	Prism angle γ		Aperture		Electrode spacing (mm)	Voltage (for λ = 520.8 nm)	
	d stage	h stage	d (mm)	h (mm)		Bias (kV)	Control (kV)
1		6'	1.4	1.4	1.6	5.3	2.1
2		12'	1.4	1.4	1.6	5.3	2.1
3		24'	1.4	1.4	1.6	5.3	2.1
4		48'	1.4	1.4	1.6	5.3	2.1
5		1°36'	1.4	1.5	1.6	5.3	2.1
6	6'		1.4	1.6	1.6	5.3	2.1
7	12'		1.4	1.7	1.6	5.3	2.1
8	24'		1.4	1.8	1.6	5.3	2.1
9	48'		1.4	1.9	1.6	5.3	2.1
10	1°36'		1.5	2.0	1.6	5.3	2.1
11		1°36'	3.0	4.0	4.5	14	6.5
12		3°11'50"	3.0	4.1	4.5	14	6.5
13		6°22'20"	3.0	4.2	4.5	14	6.5
14	1°36'		3.1	4.6	4.5	14	6.5
15	3°11'50"		3.1	4.9	4.5	14	6.5
16	6°22'20"		3.3	5.3	4.5	14	6.5
17		12°35'15"	3.6	5.6	4.5	14	6.5
18	12°35'15"		4.0	6.4	4.5	14	6.5
19[a]		−25°11'30"	4.7	7.0	6.0	20	8.8
20[a]	−25°11'30"		5.5	8.6	6.0	20	8.8

[a]These two prisms are upside down to eliminate temperature effects.
Source: Ref. 1.

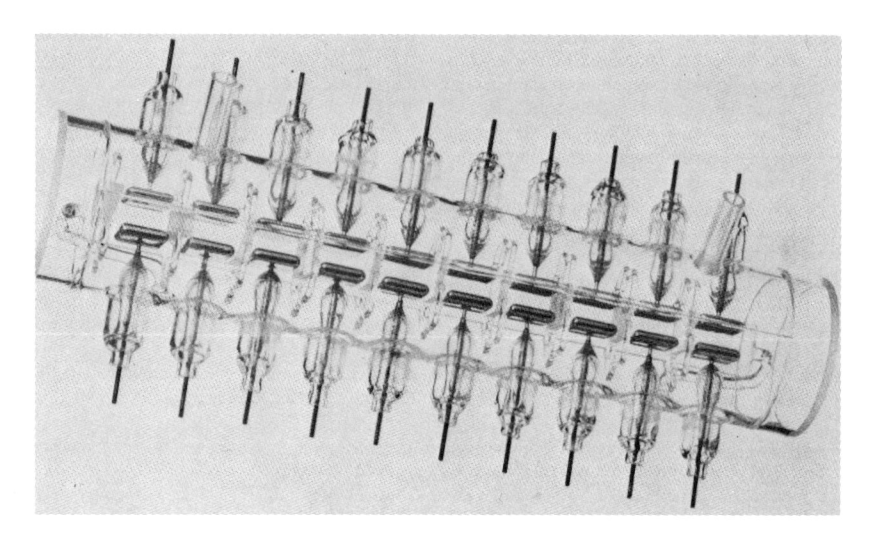

Figure 4.3 Nitrobenzene glass trough holding 10 sets of electrodes and 10 calcite prisms. (From Ref. 1.)

ture, was achieved by inverting the last two prisms. The deflection system involved very little optics but required a cell design that would maintain the nitrobenzene in an ultrapure contamination-free condition. This was achieved by an all-glass structure with glass-to-metal seals as shown in Fig. 4.3. The optic transmission for the complete system was 40% at 520.8 nm. Two-thirds of the loss was attributed to uncoated connecting optics with only one-third of the loss in the two deflector stages. The ratio of the total background light to the signal light was 0.25, which was reduced by a final polarization filter stage to 0.033. The brightest of the individual unwanted spots was 0.3% of that of the picture element.

The beam switching times obtained with the transistorized circuits employed ranged from 250 nsec at voltages of 2.5 kV to 900 nsec at 8.5 kV. To take advantage of the square-law nature of the Kerr effect, each stage was dc biased to a prerotation through 90° (see Sec. 3.1.3). To prevent the occurrence of Schlieren effects, the electrode current was kept to below 0.4 μA per stage.

There are two further important advantages that this deflector has that must be mentioned. First, the polarization switches are optically isotropic and have large angular apertures; consequently they will operate at more than one wavelength, although the drive voltages must be altered to suit. Second, they do not suffer from piezoelectric resonances as do most crystal electro-optic polarization switches.

4.1.3 Pockels Cell Deflectors

The problems associated with the use of Kerr liquids, in particular with their ability to collect contamination and the high voltages required in the final stages of deflection, have encouraged other investigators to develop deflectors using solid-state electro-optic polarization switches. Unfortunately, the crystal electro-optic polarization switch has its own set of material problems which have to date prevented these deflectors from achieving the performance of the nitrobenzene Kerr deflector.

The theory of the linear electro-optic effect and its application to light modulation is discussed in Chap. 1. In choosing an electro-optic modulator for a deflection system the following design details must be considered:

1. Polarization ratio required for each stage
2. Angular aperture required for each stage
3. Linear aperture required for each stage
4. Capacitance of stage and dielectric loss
5. Voltage required after consideration of 1 to 4 above
6. Drive power required for chosen address rate
7. Effect of piezoelectric resonances on picture definition
8. Operation for single or multiple wavelengths
9. Availability of chosen electro-optic material

Longitudinal XDP modulators have been fabricated using transparent conducting electrodes evaporated on thin Z-cut plates. This is illustrated in Fig. 4.1b. They have the advantage that large diameters and angular apertures are possible without increasing the voltage required to drive subsequent stages.

For KD*P, a 1-mm-thick Z-cut longitudinal modulator using transparent electrodes would have a half-wave voltage of approximately 3.5 kV at 632.8 nm. The capacitance would be 40 pF/cm^2 aperture. The half-wave voltage would be independent of the aperture of the cell, and a deflector using these modulators would have the important advantage that the switching voltage would be identical for all stages.

Z-cut KD*P modulators using transparent electrodes were used by Pepperl [3] at Philips GmbH Forschungslaboratorium Hamburg in a prototype three-stage DLD. Each calcite prism was combined with a glass prism to form a deflection element with parallel input-output faces. The refractive index of the glass was chosen to give symmetrical deflection of light with respect to the system axis, and all components were antireflection-coated to reduce reflection losses. The device is shown in Fig. 4.4

It seems likely that some difficulty was encountered in the development of a suitable transparent conducting electrode. Both tan δ and

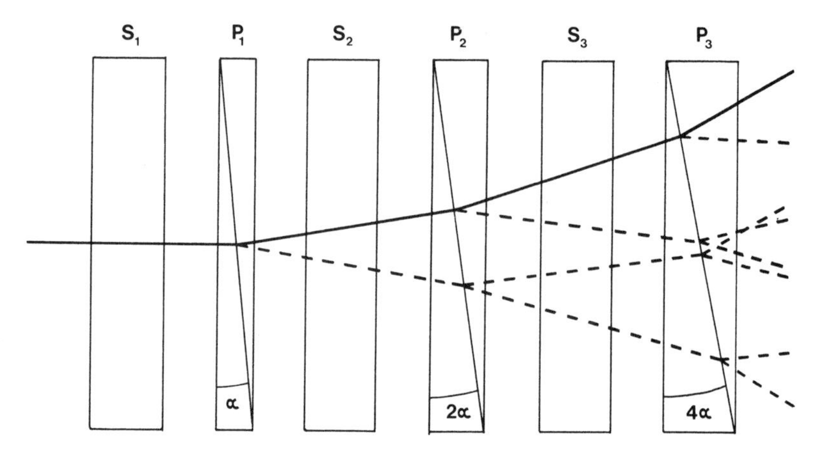

Figure 4.4 Optical arrangement for a three-stage longitudinal KD*P line deflector. (From Ref. 3.)

electrode ohmic resistance limited the maximum drive frequency to 35 kHz (see Chap. 3). This limitation occurs at the onset of thermally induced optical strain and results in a deterioration of the polarization ratio. Furthermore, the switching frequency was limited by piezoelectric resonances. These resonances were severe as no attempt was made to achieve a degree of mechanical damping. For the 1-mm crystals fundamental shear and longitudinal modes occurred at 80 and 274 kHz. The exceptionally slow access time was attributed to the driver output impedance. This gave a 30- to 50-μsec charging time for a cell capacitance of 20 pF. The optical loss of each polarization cell was given as 2%, and it would be reasonable to assume that the optic loss of each composite birefringent element was about the same value. For a 20-stage deflector the transmission losses would therefore have been considerably higher than those of Schmidt's nitrobenzene deflector.

Transverse modulators using XDP materials require crystal pairs of precise orientation and of identical optic length. Exceptional skills are required in their preparation, and their delicate nature makes this optic work even more difficult. Special mounting techniques are necessary to maintain their relative orientation and to minimize strain and maximize piezoelectric damping. It has been suggested that the only problem not present is the purification requirement pertaining to the liquid Kerr cell, and yet if devices are required to have lifetimes acceptable for commercial applications, then the crystal grower must achieve similar purity levels.

At the present time there are a number of transverse modulators that could be used for DLDs:

1. Z-cut KD*P [4]
   a. Two-crystal composite without half-wave plate
   b. Two-crystal composite with half-wave plate
2. 45° Y-cut ADP [5]
   a. Two-crystal composite with half-wave plate
   b. Four-crystal composite without half-wave plate
3. Z-cut transverse [6]
   a. Single-crystal lithium niobate

For transverse electro-optic modulators we have seen that the half-wave switching voltage $V_{\tau/2}$ is proportional to d/L, where d is the distance between the electrodes and L is the composite crystal length. For a line deflector, the beam can be deflected parallel to the electrodes and the aperture increased in that direction without increasing the electrode separation. This can be seen by inspection of Fig. 4.1d. This argument also applies to the liquid Kerr cell. Increasing the composite crystal length L reduces the drive voltage, but a practical limit is reached when either the physical aspect ratio is too small for the deflected beams or the angular aperture defined by an acceptable polarization is exceeded.

In general, transverse modulators employing half-wave plates have greater angular acceptance angles than those without. Also, transverse modulators with half-wave plates show interference figures between crossed polarizers of twofold symmetry rather than four, and this can be used with advantage to achieve large acceptance angles for line deflectors.

A line deflector using composite 45° Z-cut KD*P modulators with half-wave plates was described by Hepner [7]. This deflector, which is illustrated in Fig. 4.5, gave a single line scan of 64 positions, having six stages, with an input modulator and an output polarization filter stage. The system was immersed in an index-matching oil of refractive index 1.49, and the total optic losses for the 34 optical components used was 50% at the operating wavelength of 632.8 nm. An excellent cross-path performance was achieved for this design with a relative intensity of false positions being about 10^{-4}. The input beam diameter chosen set the electrode spacing at 2.5 mm. The calcite single prisms were cut to give 10 resolution angles per deflection position, and for the 64 deflection positions, the last deflector gave an acceptable polarization ratio for an angular field approaching 10°. The KD*P crystals each of 11-mm length required 880 V to half-wave-switch at 632.8 nm. The maximum switching frequency achieved was 100 kHz with a random access time of 10 μsec. No mention of piezoelectric resonances was made, although they were undoubtedly present. However, the resonances may have been effectively damped by the index-matching oil.

The problems associated with the fabrication of transverse modulators prevented the further development of these deflectors. More recently, improvements have been made in the design and fabrication of

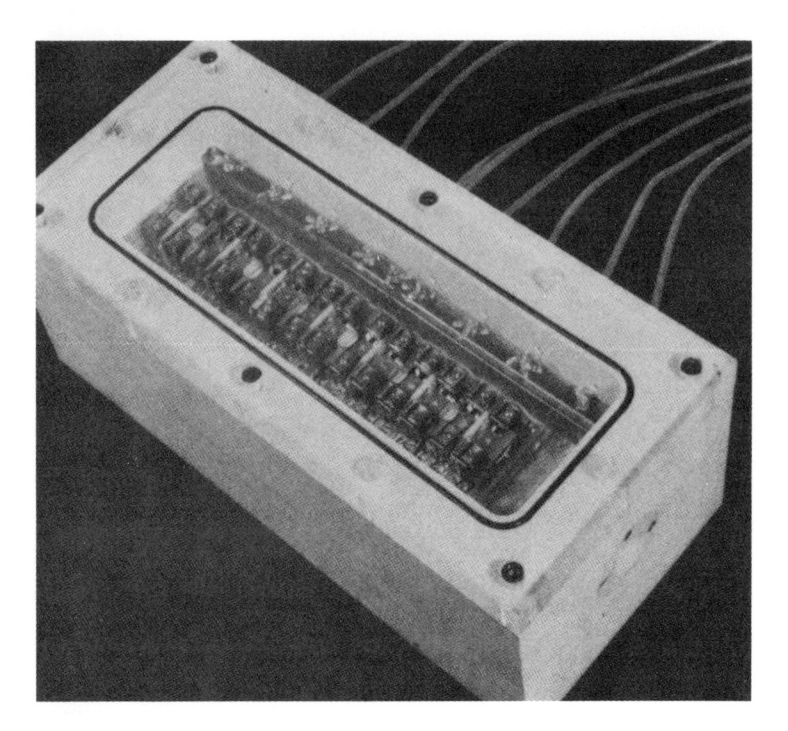

Figure 4.5 View of six-stage transverse KD*P line deflector. (From Ref. 7.)

these devices, and it is now considered that it would be possible to build useful solid-state electro-optic deflector systems. Certainly, transverse modulators of the 45° Z-cut KD*P type can now be made to a higher optic standard simply because of 10 years further manufacturing experience, and improvements have obviously occurred in the performance of high-voltage drive circuits. Consequently, switching rates of 1 to 10 MHz are now possible with access times of less than 0.1 μsec at polarization ratios of $1{:}10^4$ or more [8]. It would also be possible to build a deflector using 45° Y-cut transverse modulators, and these would be free from piezoelectric resonances.

The development of new electro-optic crystals with improved characteristics could change this situation. The availability of a cubic optically isotropic crystal, capable of producing a modulator requiring a drive voltage of a few hundred volts or lower, would most certainly revive interest in these potentially useful devices.

4.2 ANALOG BEAM DEFLECTORS

4.2.1 Device Designs

For both prism and refractive index gradient deflectors the number of resolvable beam positions obtained is proportional to the induced refractive index change Δn, which is related to the electro-optic coefficient r_{ij} and the third power of refractive index by

$$\Delta n \propto r_{ij} n^3 = M_R \tag{4.1}$$

M_R was used by Thomas [9] as the figure of merit for spatial resolution. Also, as the speed of deflection is for many applications as important as the number of resolvable beam positions and since the speed of operation depends on the device capacitance, a material figure of merit for temporal resolution is given by

$$M_\tau = \frac{r_{ij} n^3}{\varepsilon_{ij}} \tag{4.2}$$

Figures of merit are listed in Table 3.1 and can be used as a basis for material selection. The performance of a particular deflector can be optimized by choice of crystal orientation and dimensions and by careful design of the electronic drive circuits.

Although ADP is the least attractive electro-optic material, judged by the figures of merit M_R and M_τ given in Table 3.1, it has some advantages and has been used by some workers for constructing analog deflectors.

It has been shown by Magdich [10] that the 45° Y-cut ADP crystal is nonpiezoelectric, and it is for this reason that it has become the most widely used material for low-voltage light modulators. It has found many industrial applications where lasers require amplitude control, and an example of a six-channel 45° Y-cut ADP light modulator is shown in Fig. 4.6. This modulator requires 300-V drive at 488 nm to give full amplitude control, and the modulated light is resonance free. It is for these reasons that ADP was chosen by Beasley [11] for his deflector in an experimental large-screen TV projection display. The electro-optic multiple-prism scanner he constructed is shown in Fig. 4.7. The scanner had an aperture of 40 × 2.5 mm and therefore required only 10 prisms of 80-mm base length to give a light path of 400 mm. The price paid for the relatively few number of prisms involved was a high capacitance (3000 pF). The line scan at 60 Hz needed 14-kV drive to give 160 resolvable spots for single-pass operation. The field strength of 2.8 MV/m could not be sustained without electrical breakdown, and it is probable for this range of ionic crystals that this

Figure 4.6 Six-spot ADP modulator.

field strength is too high. An electro-optic modulator of ADP would today be given a maximum permitted operational field strength of 0.5 MV/m. An alternative ADP multiple-prism design used by Thomas [12] was built by Coherent Associates (USA) and had an aperture of 2 × 2 mm and an overall crystal length of 200 mm. This deflector was evaluated for use in a crystal streak camera. The maximum recommended voltage of ±2 kV reduced the field strength to 1 MV/m and no doubt improved the lifetime. This streak camera required low-repetition-rate impulses of between 1- and 100-nsec rise time, which would not have overstressed the material, and it is probable in this application that this deflector would have had an acceptable lifetime. The capacitance of 100 pF prevented very fast switching, and the circuits employed achieved a 20-nsec sweep, giving a temporal resolution of approximately 2.5 nsec per resolvable spot at 1.06 μm. A krytron sweep circuit [13] would have improved this resolution to possibly 500 psec, but this is still two orders of magnitude slower than currently available with image convertor streak cameras [14]. It therefore seems unlikely that ADP would produce anything other than an experimental deflector for this application.

We have seen that the M_R and M_τ values for $LiNbO_3$ indicate that it is more suited to analog deflection than ADP. For analog deflectors

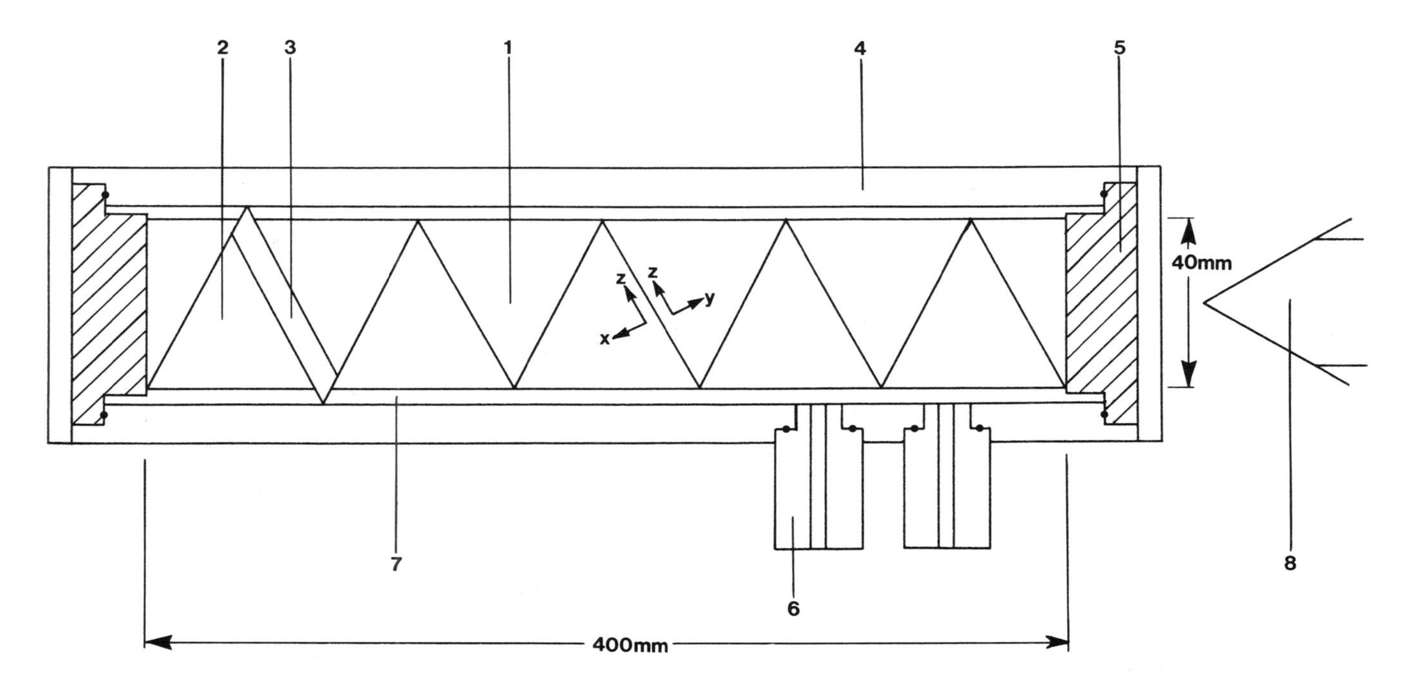

Figure 4.7 Bulk multiple-prism line deflector. (1) Prism in main deflector section; (2) prism in third harmonic deflector section; (3) glass spacer; (4) outer case; (5) end windows with O-ring seals; (6) high voltage connector; (7) Teflon spacer; (8) polarized light beam. (From Ref. 11.)

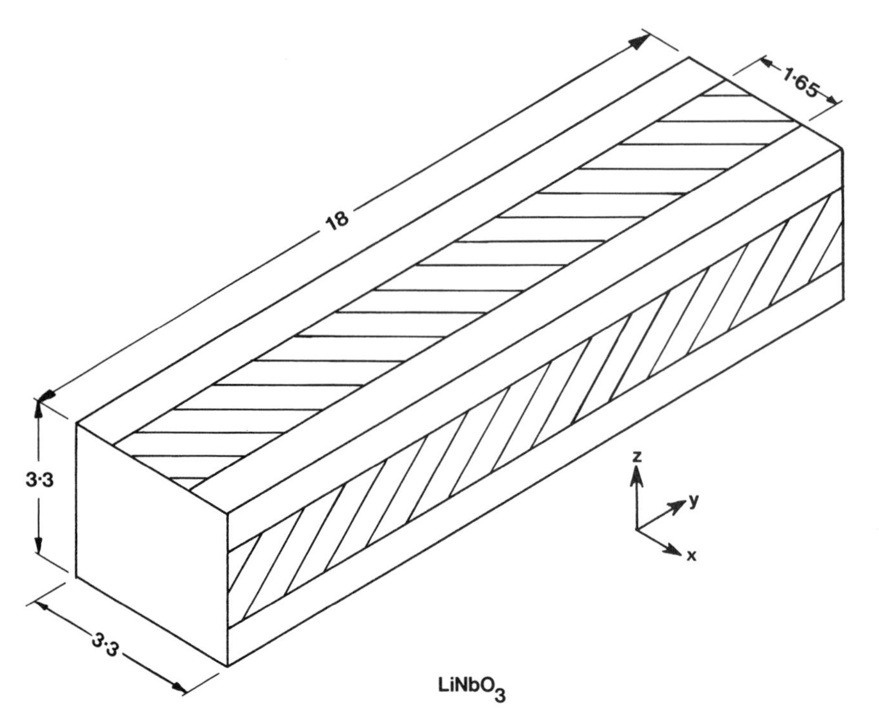

Figure 4.8 Quadrupole deflector using planar electrodes. (From Ref. 15.)

using $LiNbO_3$ it is useful to apply a field parallel to the z axis to obtain a refractive index change $\Delta n = (1/2)n_e^3 r_{33} E_z$ for light polarized in the direction of the applied field. The refractive index change is four times that obtained from ADP when the field is applied in the y direction and for which

$$\Delta n = \frac{\sqrt{2}\, n_o^3 n_e^3 r_{41} E_y}{(n_o^2 + n_e^2)^{3/2}} \tag{4.3}$$

Figure 4.8 shows an $LiNbO_3$ quadrupole deflector of dimensions 3.3 × 3.3 × 18 mm built by Gisin et al. [15]. Electrodes of width equal to half the face width were deposited on each of the side faces to produce a gradient field within the crystal. As the deflector was used for on-off light modulation by beam displacement from a detector, high resolution and a linear deflection were not necessary, and the simple electrode geometry sufficed. Nevertheless, it was stated that the deflection increased linearly with the voltage throughout the investigated range of −2.8 to +2.8 kV. The electrode geometry gave the device a low capacitance, and for a 3-mm beam at 632.8 nm the drive voltage required would have been approximately 200 V. This is in itself a reasonable specification for a modulator. The on-off modulation that the deflector provided at approximately 1 MHz did not, apparently, suffer piezoelectric resonance effects, but the techniques employed to achieve the necessary damping were not described. The $LiNbO_3$ gradient deflector was further developed by Ireland [16-18] using a cylindrical electrode geometry to produce a linear field gradient. The crystal used was a 5 × 5 × 32 mm Y-cut rotated 45° to the z axis. This configuration produces, to a close approximation, a linear field gradient in the x and z directions. In the second publication an aperture defined by the hyperbola xy = ±1, where x and y are in millimeters was chosen and examined theoretically to see how *best fit* cylindrical electrodes would degrade its operation. Ireland showed that in using cylindrical electrodes of radius 2 mm the field error at $N_R = 40$ degraded the diffraction-limited condition by only a few percent, this degradation being less than that produced from a ±12-μm positional tolerance on the electrodes. Each of the 2-mm-radius electrodes were stopped 2 mm away from the polished end faces of the crystal to avoid surface breakdown problems. Additionally, the tracking distance around the "arrowheads" (see Fig. 4.9) was 3.4 mm so that the device would withstand a voltage of ±8 kV when immersed in a high dielectric fluorocarbon fluid which was used to reduce the surface reflection losses.

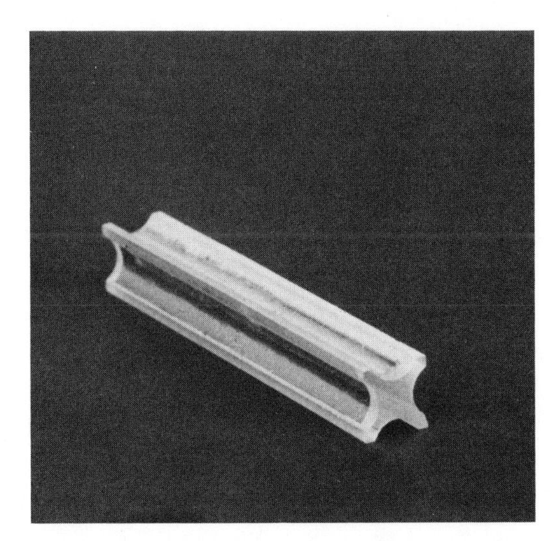

Figure 4.9 Lithium niobate crystal fabricated and used by Ireland and Ley in crystal streak camera experiments. Crystal dimensions, 5 × 5 × 32 mm.

Static optic tests were performed with the deflector using a collimated beam at 632.8 nm from an He-Ne laser. This beam was magnified and spatially filtered so that the deflector aperture was uniformly illuminated. A 50-μm slit was positioned parallel to the x axis and immediately after the deflector. By using a 50-cm-focal-length lens, the far-field spatial distribution was examined with voltages in the range 0 to 5 kV applied to the device. The distributions were very close to being *diffraction limited* over the whole voltage range. Photographic records showed that the main difference between the distributions and the $(\sin x/x)^2$ function expected for perfect diffraction at a slit was an increase in the intensity of the subsidiary diffraction maxima. For example, at 5 kV the first such maximum was 8% of the peak intensity rather than 4.7%. This "spreading" of energy is consistent with a wavefront aberration of $\lambda/2$ [19]. The deflector produced a static deflection that was a linear function of voltage to within better than 2%. It required 100 V to produce one resolvable spot at 632.8 nm. By using (3.42) with $r_{33} = 30.8 \times 10^{-12}$ mV^{-1} and $n_e = 2.20$, this corresponds to an E_z(max) of 4.5 MV/m at 5 kV. Electrical tests on the deflector showed that it had a capacitance of 45 pF, so that driven by two 50-Ω lines it had a potential rise time (10 to 90% amplitude) of ~2.5 nsec. TDR measurements with a 30-psec resolution Tektronix type 7S12 pulse sampling unit confirmed this result.

For operation in the streak mode the deflector was driven by a pair of 50-Ω lines from an EG and G KN22B Krytron switch. With the transmission lines and the Krytron envelope carefully screened, an 8-kV voltage ramp could be applied to the deflector in 4.0 nsec. By backing off a pair of the deflector electrodes to −4 kV, the interelectrode pd was swept from +4 to −4 kV when the Krytron conducted. This allowed the beam to be swept symmetrically through the zero field (low-aberration) position. To verify the deflector could achieve the temporal resolution and dynamic range implied by the optic and electrical tests outlined above, it was used in conjunction with frequency-doubled pulses from a mode-locked Nd-YAG laser. These pulses had previously been measured with an IMACON 675 image convertor streak camera (resolution ~5 psec) and found to be of 27-psec duration. Before entering the deflector, these 0.53-μm pulses were passed through a spatial filter and 2.0-cm BK-7 etalon with 70% reflective coatings. The etalon was used to calibrate the streak records. It produced multiple pulses, temporally separated by 200 psec, and of 2:1 intensity ratio. Figure 4.10a shows a microdensitometer scan of a far-field streak record. It can be seen that the recorded FWHM intensity width for the pulses is 50 psec. Assuming a quadratic temporal folding, this implies that the camera resolution was 42 psec. By comparing this with earlier records, Ireland showed that the use of the slit immediately after the deflector led to a considerable improvement in the dynamic range of the camera. These results indicated that satellite pulses of 5% of the main pulse intensity could be detected.

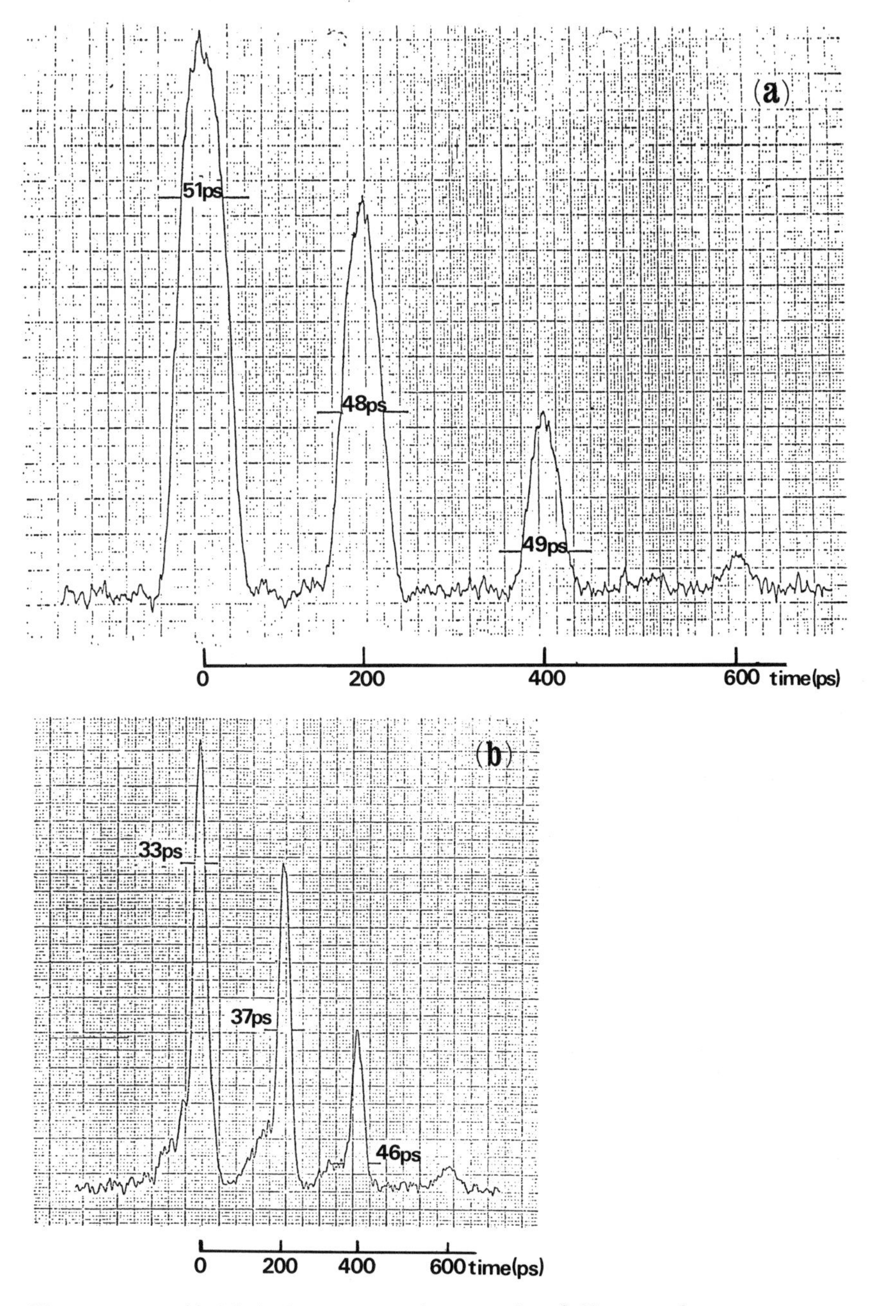

Figure 4.10 Multiple image streak records of 27-psec frequency-doubled Nd-YAG laser pulse. (a) Quadrupole deflector, single pass; (b) quadrupole deflector, double pass.

To achieve better temporal resolution, the camera was set up with the deflector double-passed. In this case the 50-μm slit was positioned immediately before the deflector and a 100% reflectivity mirror after it. A partially reflecting mirror in the beam was used to deflect the double-passed pulses for focusing and recording. Figure 4.10b gives the microdensitometer record obtained with this experimental arrangement. From this and other records it was found that double-passing the deflector improved the resolution to 20 psec as expected while adversely affecting the dynamic range.

4.2.2 Comparisons Between Multiple-Prism and Gradient Deflectors

From an examination of the equations for both multiple-prism and gradient deflectors it can be concluded that for devices of identical aperture and crystal length the deflection angle for a given value of Δn will be the same; that is,

$$\Delta\phi = 2L\,\frac{\Delta n}{Wn_2} \qquad (3.26)/(3.41)$$

In calculating n for a further comparison, we shall assume identical-length deflectors of the same aperture and of diameter d, as shown in Fig. 4.11.

Obviously the field strength for case a is V_{app}/d, and for case b, using $zx = d^2/8$, the field strength at the aperture edge is $2V_{app}/d$. The gradient deflector will therefore give twice the deflection for the same aperture since $\Delta n \alpha V_{app}/d$.

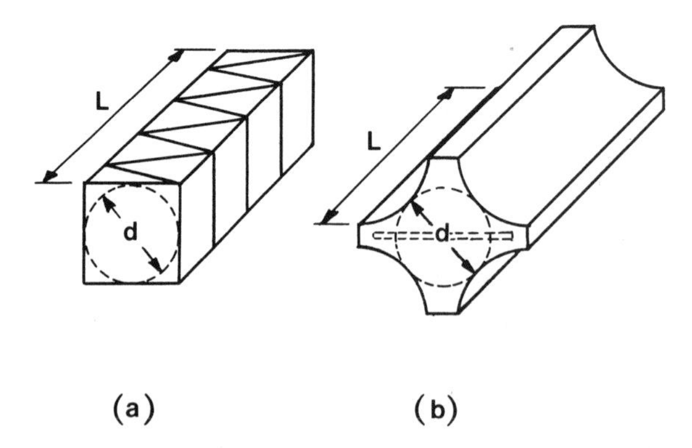

Figure 4.11 Analog prism deflectors. (a) Multiple prism, (b) gradient quadrupole.

It is interesting to make a direct comparison for one possible E-O material, and for this purpose AMO has been chosen, since it has the highest figures of merit of known E-O materials and is potentially the most useful (see Table 3.1).

By taking Thomas' information from Table 3.1,

$$r_{52} = 327 \times 10^{-12} \text{ m/V}$$

$$n \simeq 1.5$$

$$\varepsilon \simeq 17$$

and by choosing

$$a = d = W = 2 \times 10^{-3} \text{ m}$$

$$L = 60 \text{ mm}$$

$$V_{app} = 6 \times 10^{3} \text{V}$$

and then using (3.26),

$$\Delta\phi = \frac{2L}{a}\frac{1}{2}n^3 r_{52}E$$

we find for the multiple-prism deflector, $\Delta\phi$ = 96 mrad, N_R = 150, and for the gradient deflector, $\Delta\phi$ = 192 mrad, N_R = 300.

The situation can be further improved in favor of the gradient deflector by extending the beam width using a slit aperture, as indicated in Fig. 4.11b. This will lead to an improvement in effective aperture by utilizing the gradient field well into the arrowheads of the crystal. (This technique was used by Ireland [17] for his crystal streak camera, as maximum resolution was required only in the deflection plane.) Such large deflection angles would require beam focusing to prevent vignetting, but to date deflectors have not been constructed where vignetting has required this preventative measure.

It is also necessary to examine the performance of both deflectors as crystal dimensions are altered. In Chap. 3 it has been shown that

$$N_R = \frac{2L\,\Delta n}{\varepsilon\lambda} \qquad (3.27)/(3.42)$$

For a fully filled aperture the spatial resolution is proportional to the crystal length and to the induced refractive index change but is independent of aperture. The value of Δn, however, is inversely proportional to the aperture for both deflectors, and, surprisingly, for

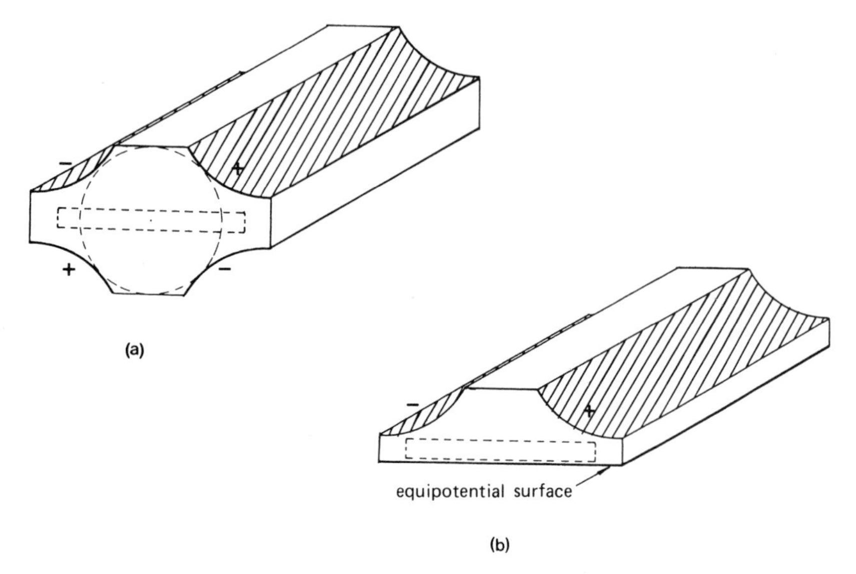

Figure 4.12 Variation to the quadrupole design. (a) Truncated quadrupole deflector, (b) optimized gradient deflector.

a given drive voltage, small apparently means good. There are unfortunately several physical limitations which prevent this conclusion from being realized in practice. First, it is extremely difficult to manufacture from fragile electro-optic crystals optic components when dimensions are reduced to below 2 mm. Second, vignetting occurs for quite moderate deflection angles in this situation. For example with a 2-mm aperture, L = 60 mm and ϕ = 96 mrad, an impractical situation results for which prefocusing is no longer a solution. Third, and again taking the example of a 2-mm aperture, the electrode spacing gives for the multiple-prism and gradient deflectors field strengths of 3 and 6 MV/m, respectively. This is close to the known breakdown strength of the XDP range of materials [20] and could well lead to refractive index damage in the oxygen octahedra ferroelectrics. Refractive index damage believed to be of this type has recently been observed in $LiNbO_3$ by the authors.

Finally, although N_R increases linearly as the aperture is reduced, the device capacitance also increases with the same relationship; therefore, for a given voltage source the temporal resolution remains constant and is not affected by device aperture. For a display system this may be an acceptable price to pay for an increased deflection angle, but for a crystal streak camera no advantage would be gained.

Within these constraints there are some possible variations. The aperture of the multiple-prism deflector can be widened in the plane of the

Table 4.2 Analog Deflector Characteristics Taken from Reviewed Publications

Deflector type	Reference	Application	Material dimensions (mm)	Number of resolvable spots (N_R) wavelength (mm)	Deflection per kilovolt (mrad/kV)	Voltage drive (kV), rise time (nsec)	Capacitance (pF)
9 multiple prisms	J. Beasley	TV line scan	ADP: 2.5 × 40 × 400	160 568	0.33	±7	3000
Multiple prisms	S. W. Thomas	Streak camera	ADP: 2.0 × 2.0 × 200	18 532	1.3	±2 20	100
Quadrupole cylindrical electrodes	C. L. M. Ireland	Streak camera	$LiNbO_3$: 5 × 5 × 32	100 632.8	1.8	±4 4.0	45
20 multiple prisms (proposed)	S. W. Thomas	Streak camera	AMO: 2 × 2 × 60	153 1060	16	±3 0.5	10
Quadrupole plane electrodes	B. V. Gisin	Light modulator	$LiNbO_3$: 3.3 × 3.3 × 18	42 632.8	1.85	±2.8	16
Quadrupole cylindrical electrodes (proposed)	Text example	Streak camera	AMO: 3 × 3 × 60	300 1060	32	±3	40 unshaped 10 shaped

electrodes [11], and this reduces the prism elements required and eases the manufacturing difficulties. The spatial resolution is unaffected, and the capacitance is increased, but for display systems this could be tolerated. If a slit aperture is used, the gradient deflector can be trimmed to a minimum capacitance value by removing material no longer required, as shown in Fig. 4.12a. This truncated design can reduce the capacitance by a factor of 2. A further reduction of 2 can be obtained if the device is reshaped as shown in Fig. 4.12b. The aperture is further restricted in height but not in width, and for a streak camera application this could be acceptable. By these two methods the capacitance of the gradient deflector could be reduced to that of the equivalent aperture prism deflector while retaining its resolution advantage. (See Table 4.2.)

4.3 ANALOG-DIGITAL ARRAY DEFLECTORS

4.3.1 Bulk Crystal Deflectors

The concept of an array of prism deflectors was introduced in a paper by Ninomiya in 1973 [21] and further expanded by the same author in 1974 [22]. Both of these papers dealt with bulk prism array deflectors, and it is significant that they were developed for the possible application to optic data processing and optic memory systems. For these applications work in integrated optics has now led to the development of a range of deflectors using the array concept of Ninomiya but involving surface-induced electro-optic refractive index effects rather than the bulk effect discussed so far.

The basic electro-optic prism array deflector is shown in Fig. 4.13a, where the type B prisms are used for index-matching purposes only. This basic design has the advantage that it can be built on a single-crystal slab without interprism spacing, provided that the deflector is used in a digital mode for main diffracted orders only. Nevertheless, this simple arrangement does give the closest approach to a sawtooth phase grating and will, at the main diffracted order positions, give a high signal to noise ratio.

For ease of manufacture, Ninomiya [21] described a deflector manufactured from a single plate of Z-cut lithium niobate, as shown in Fig. 4.13b.

To achieve continuous deflection, adjacent prisms were electrically isolated so that bias voltages could be separately applied. However, the discontinuities produced by these isolation spaces resulted in a poor signal-to-noise ratio, and because of this, Ninomiya considered this slab design to be an early prototype. He then proposed an alternative design of a 20-prism array of oppositely orientated prisms not requiring interprism spacing and capable of giving 630 resolvable positions for a drive voltage of ±800 V. (See Table 4.3.)

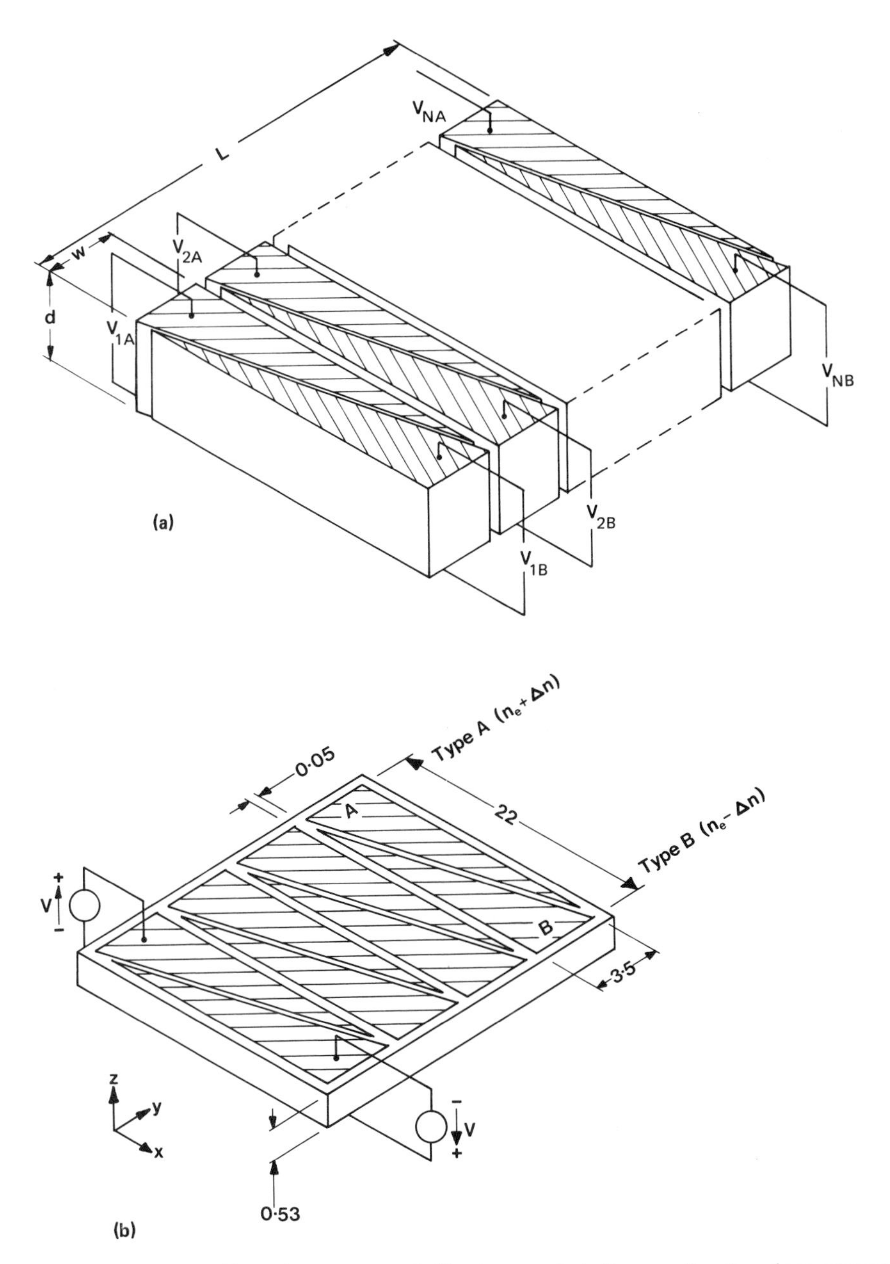

Figure 4.13 Bulk prism array deflectors. (a) Generalized prism array deflector, (b) lithium niobate slab deflector [21], (c) nine-crystal $LiNbO_3$ prism array deflector [22], (d) nonlinear field in prism deflector.

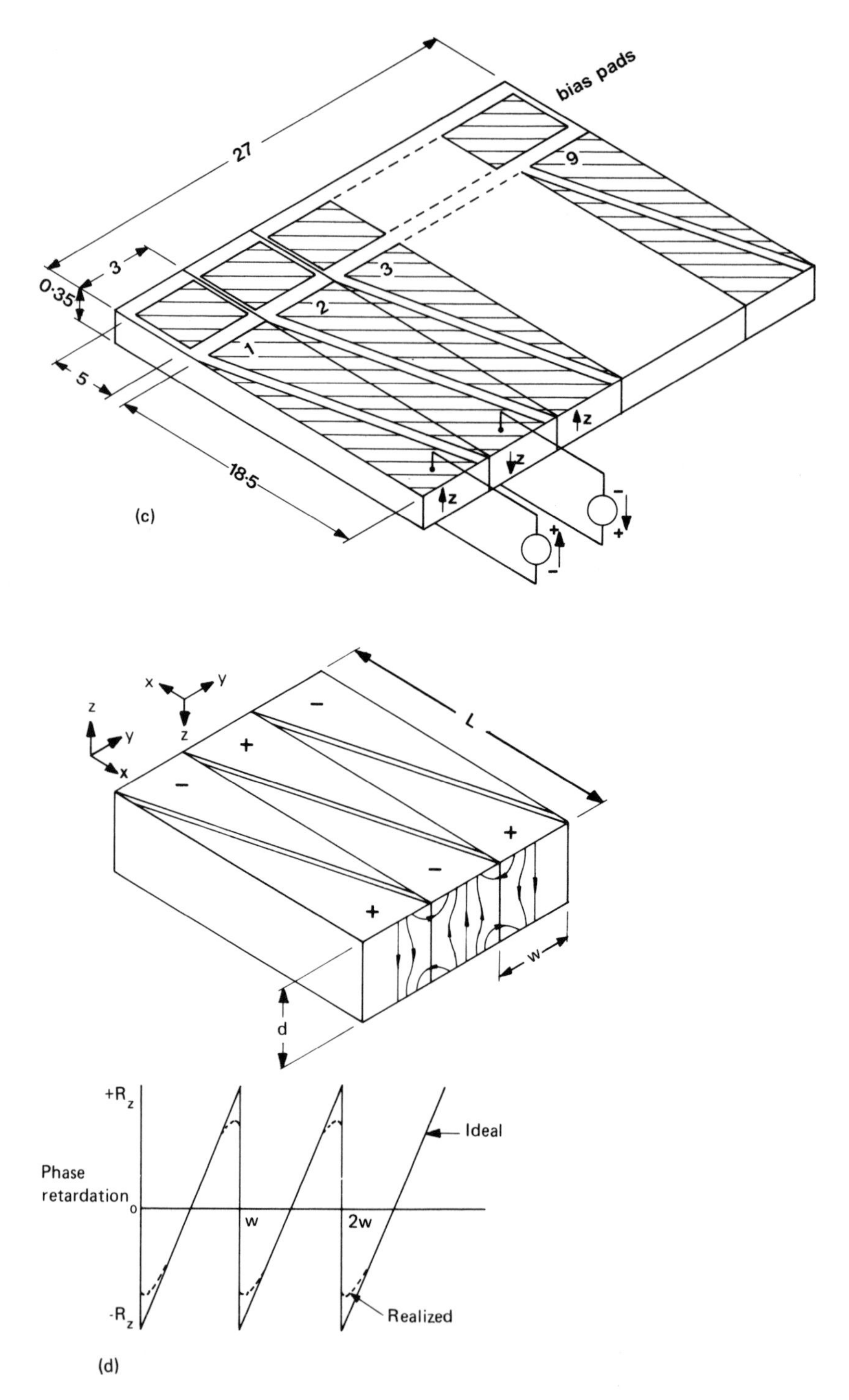

Figure 4.13 (Continued)

Table 4.3 Bulk Array Deflector Characteristics[a]

Deflector	Resolvable positions (analog drive), N_R	Dimensions of single prism (mm)			$V_{\lambda/2}$ (V)	V applied (V)	C (pF)	V_{app}/N_R	Material
		w	d	ℓ					
4-section array slab design	50	3.5	0.53	23.5	48	±597	140	22	$LiNbO_3$
20-section array individual prisms (proposed)	630	2	0.5	25	40	±800	500	2.5	$LiNbO_3$
9-section array individual prisms	180	3	0.35	18.5	38.5	±580	358	6.5	$LiNbO_3$
54-section array individual prisms (proposed)	1080	0.8	0.2	21	16.4	±164	1750	0.34	$LiTaO_3$

[a]*Note:* All data given for 632.8 nm.
Source: From Refs. 21 and 22.

Ninomiya's paper of 1974 gave details of a nine-prism array of the above design, shown in Fig. 4.13c. This design allowed adjacent prisms to be joined electrically with both type A and B prisms operating electro-optically. It did, however, preclude the application of individual bias voltages to the prisms, and to overcome this problem, bias pads were added.

Although this design did give an improved signal-to-noise ratio, sidelobes still occurred at the main diffraction order positions owing to a fringing field effect at the regions of inverted polarity. This is illustrated in Fig. 4.13d.

Ninomiya calculated the field across the aperture (w) for a line midway between the electrodes to take into account the fringing field effect. In these calculations he defined the effective thicknesses as

$$p = \frac{\varepsilon_x}{\varepsilon_z} \frac{d}{w} \tag{4.4}$$

For the nine-prism array constructed, p was equal to 0.28, and the calculations indicated a respectable sawtooth wavefront. By using a definition for signal-to-noise ratio of

$$S/N = \max_V \left(\frac{I_0}{\max\{I_+, I_-\}} \right) \tag{4.5}$$

where $\max_V$ means the maximum of the function V and I_0 is the intensity of the main lobe and I_+ and I_- are the intensities of the plus and minus first-order sidelobes, respectively, these calculations gave the signal-to-noise ratio as 10:1 for the tenth order of diffraction. In his experimental work the deflector achieved a slightly better signal-to-noise ratio than his calculations indicated. Finally Ninomiya proposed a 54-crystal 1080-spot deflector using $LiTaO_3$, but it is not known whether this device was built.

4.3.2 Surface-Wave E-O Deflectors

To take maximum advantage of the electro-optic effect, by using the principles discussed for single-crystal prism deflectors, it is necessary (for a given beam width) either to increase the interaction length L or to increase Δn by increasing the electric field strength [see (3.27)].

In practice it is more difficult to double a crystal length than to halve its thickness. With bulk deflectors the tendency has been to increase Δn by reducing aperture height, with a geometric limit set on this from a maximum acceptable diffraction condition and, also, with a practical limit set by fabrication and polishing constraints. Consequently bulk prism deflectors of moderate resolution have been built with slit apertures of 0.5 mm in height, and any further attempt to reduce this to 0.1 mm has been frustrated.

There is therefore a good argument for building a surface waveguide deflector if, as a result, higher field strengths can be achieved. We have also seen that reducing the aperture width W for a single-prism deflector leaves the resolution unaltered but that again fabrication difficulties prevent very narrow crystals from being produced for array deflectors. In surface waveguide deflectors using two-dimensional electrode systems the possibility exists that novel designs would allow a large number of narrow-aperture prisms to be arrayed as an integrated component. Such a device would apparently achieve the ideal conditions for E-O beam deflection, as high field strength, reasonable interactional length, and simple array configurations are possible.

The hope of designing a deflector with an improved resolution and the attractive commercial possibility of developing a device that could form a critical part of many integrated waveguide systems has resulted in the recent publication of a number of novel electro-optic surface-wave prism deflector (ESP) devices.

A promising planar configuration device based on a Y-cut diffused $LiNbO_3$ waveguide was devised independently by Kaminow and Stulz [23] and Tsai and Saunier [24]. The basic deflector is formed by narrow electrodes deposited on the surface of an optic waveguide. Two electrodes are parallel to the direction of beam propagation, and the third is tilted to give a surface field gradient. This effectively simulates a thin-film electro-optic prism deflector. The electrode arrangement is shown in Fig. 4.14a.

An example of an uncompensated $LiNbO_3$ waveguide array structure built by Tsai and Saunier with A = 150 μm, L = 10 mm, and N = 4 required 8 V per beam position. These were completely resolved at 632.8 nm. With a capacitance of 4 pF the device had a subnanosecond switching capability or as a modulator, driven from a 50-Ω source, a base bandwidth of 1.66 GHz.†

An alternative approach by Sasaki [25] and Saunier et al. [26] shown in Fig. 4.15 uses an array of interdigital finger electrodes to give a linear variation of phase retardation across an array of waveguide channels.

---

†*Note:* The lithium niobate surface waveguide was formed by diffusing a layer of titanium approximately 10-Å thick into the surface of Y-cut $LiNbO_3$. This required heating to approximately 980°C in an inert atmosphere for a period of 4 hr [25].

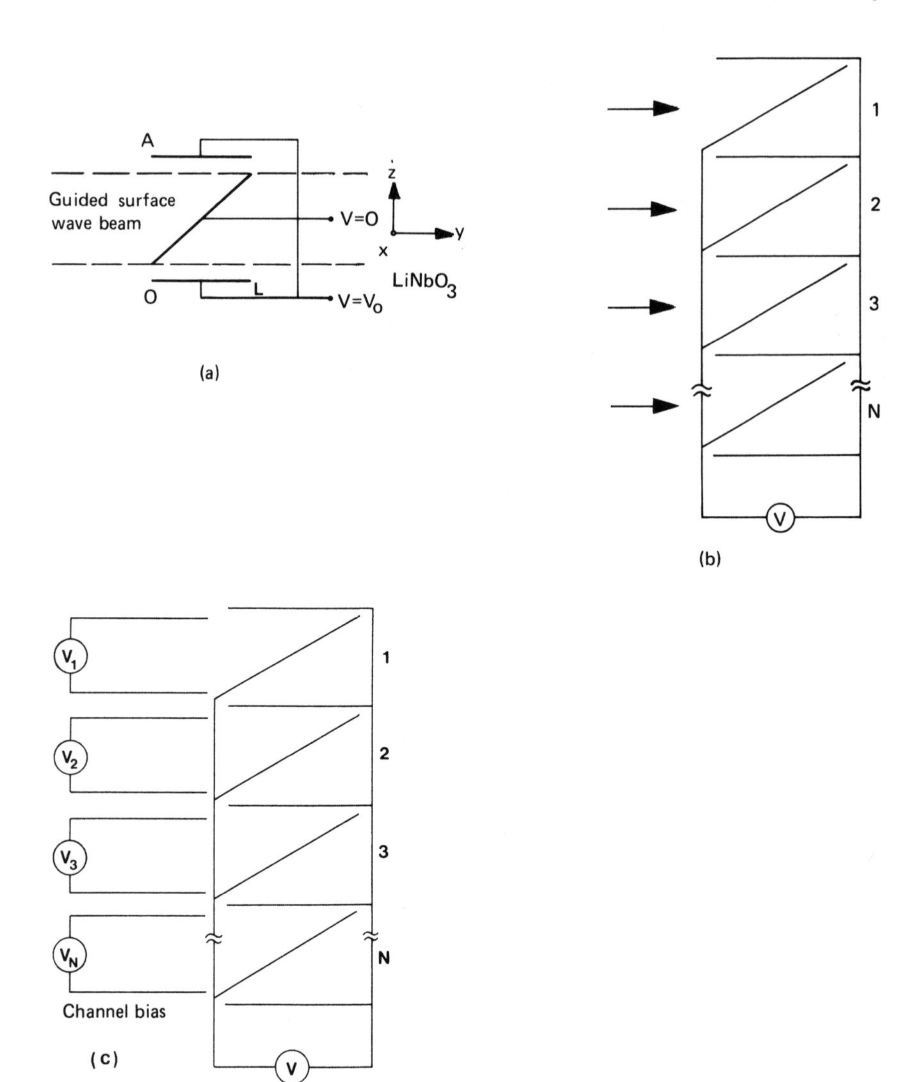

Figure 4.14 Schematic patterns for ESP devices. (a) Single-prism element, (b) array of prism elements, (c) surface array with phase linearization.

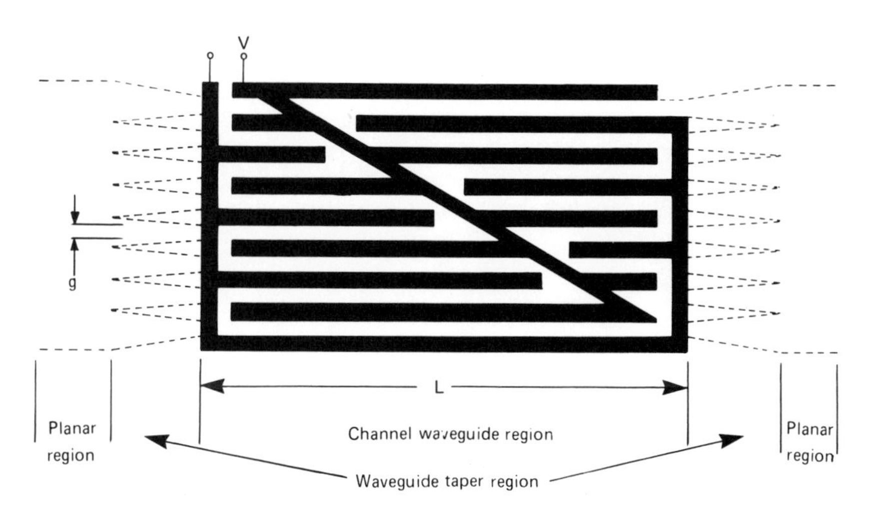

Figure 4.15 Schematic multichannel waveguide deflector.

The electro-optically induced phase for light entering the device at aperture coordinate z is given by

$$n(Z) = \frac{-\pi n_e^3 r_{33}}{\lambda_0} \int_0^L E_z zy \, dy \tag{4.6}$$

Kaminow and Stulz have shown that the linear phase shift produced across the aperture causes the wavefront to rotate through a scan angle ϕ given by

$$\tan \phi = \frac{2\lambda_0 n_o}{\pi n_e A} \tag{4.7}$$

where

$$n_o = \frac{2n_e^3 r_{33} V_0 L}{\lambda_0 A} \tag{4.8}$$

When a number of basic prism deflectors are placed side by side along a line perpendicular to the direction of propagation of the light beam, an ESP array is produced, as shown in Fig. 4.14b. Recently, Revelli [27] has shown that the periodicity involved in arraying prisms in parallel increases the number of resolvable spots by only a factor of 2 on that of a single element alone. To overcome this difficulty, a surface-wave analog to Ninomiya's [22] bulk device might be constructed in waveguide form, as indicated in Fig. 4.14c. By this bias technique the number of resolvable spots would approach the theoretical value of N times that of the single-element resolution. In practice the realization of such a solution is somewhat complicated, and no experimental details are available for phase-corrected ESP arrays.

If ϕ is the incremental phase change produced between two adjacent waveguide channels, then it has been shown by Bulmer et al. [28] that the number of spots is

$$N_R = \frac{N \Delta \psi}{\pi} \tag{4.9}$$

and that by taking the electric field midway between the electrodes as

$$E_z = \frac{2V_{app}}{\pi g} \tag{4.10}$$

where g is the spacing between electrodes, then

$$\Delta \phi = 2N_0^3 \frac{r_{33} V_{app} L}{\lambda_0 g N} \tag{4.11}$$

and the number of resolvable spots becomes

$$N_R = 2N_0^3 \frac{r_{33} V_{app} L}{\pi \lambda_0 g} \tag{4.12}$$

or the voltage required to deflect the spot to its first position is given by

$$V_1 = \frac{\pi \lambda_0 g}{n_e^3 r_{33} L} \tag{4.13}$$

The deflection is continuous, but the maximum useful deflection is through N_R resolvable spots, which is limited to the number of channels N.

For the device built by Sasaki [25] where N = 20, L = 18 mm, and g ≈ 10 μm, (4.13) gives a voltage of 1.7 V for one resolvable spot position. The experimental device achieved about 16 resolvable positions for a voltage swing of ±16 V, although the signal-to-noise ratio was poor at voltage levels above 5 resolvable spot positions. At 148 pF the capacitance of this device was high for the resolution obtained.

A similar five-channel device constructed by Saunier et al. [26] of dimensions L = 3.6 mm and g ≈ 7.5 μm required 3.1 V for one resolvable

Table 4.4 Electro-optic Bragg Surface-Wave Deflector Using $LiNbO_2$

	i	ii	iii	iv
Finger gap width, a (μm)	3	4	6	13.33
Length of fingers, ℓ (mm)	1.6	3	6.6	
Number of pairs of fingers, N	42	31	21	15
Drive voltage for maximum intensity in deflected beam, V_m	8.3	41	2.6	9.5
Deflection angle, $2\theta_B$	3.03	2.26	1.51	1.24
Capacitance, C(pF)	42	58	88	
Cutoff frequency, fc (MHz)	151	110	72	
Deflection efficiency, η (%)	96	93	94	95

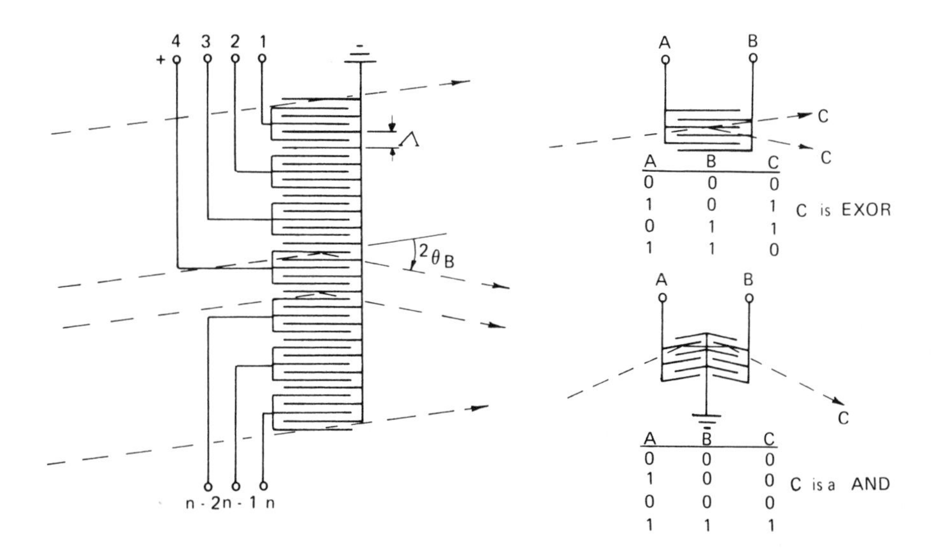

Figure 4.16 (a) E-O Bragg surface-wave spatial modulator. (b) E-O Bragg surface-wave computer logic elements.

spot position. Saunier demonstrated the recycling property of this type of deflector and operated his device up to 45.5 V finishing the fourth cycle of five resolvable positions. The capacitance was only 4.3 pF, which demonstrates the critical dependence of the type of deflector on the design of the electrode array. This device is being studied for application to gigahertz rate A/D convertors.

A Bragg deflector can be made using the E-O effect to produce a refractive index grating in an optical waveguide [29]. As with acoustic Bragg deflectors, light enters the grating region at an angle

$$\sin \theta_B = \frac{\lambda}{2\Lambda} \tag{4.14}$$

and when $Q = 4\ell\lambda/\Lambda >> 1$, that is, the Raman-Nath parameter is large, light can be reflected with high efficiency, and periodic exchange takes place between the undeflected and deflected beams with complete power transfer to the deflected beam occurring when

$$\frac{\ell\, \Delta n}{\cos \theta_B} = \frac{n\lambda_0}{2} \tag{4.15}$$

Auracher et al. [29] used Y-cut $LiNbO_3$ as the E-O substrate but produced a waveguide by out-diffusion of LiO_2 for 0.5 hr at 980°C in a flowing O_2 atmosphere. The interdigital electrode structure used was designed for a beam width of 0.5 mm, and the three electrode configurations tested are shown as deflector designs i, ii, and iii in Table 4.4. Clearly these fixed grating Bragg deflectors are digital deflectors as the deflection angle cannot be altered without changing the electrode configuration. They are, however, potentially very interesting devices, particularly as the active component for modulation and deflection in integrated optics. Already a 32-element integrated-optic spatial modulator IOSM has been built by Veber and Kenan [30] and computer logic elements proposed using Bragg E-O surface-wave deflectors. These devices are shown in Fig. 4.16. There is a variety of possible parallel and serial combinations for EOBDs (electro-optic Bragg deflectors) that would achieve multispot deflection, and many integrated devices will no doubt be built and evaluated. Nevertheless, it does seem from the devices already evaluated that this simple and efficient deflector may form the building block for future integrated-optic systems.

It is also of interest to those working in this area that a single-spot EOBD could be used as an efficient high-speed modulator, replacing many of the bulk E-O and A-O modulators now in use. For discrete component applications tapered-gap prism couplers have recently reduced insertion losses to below 0.5 dB (Sarid et al. [31]), and piezoelectric induced acoustic resonances familiar to those who have experimented with E-O $LiNbO_3$ have been successfully damped by simple mechanical means (Ramachandran [32]).

REFERENCES

1. U. J. Schmidt, *Philips Tech. Rev. 36*:117 (1976).
2. H. Meyer, D. Riekmann, K. P. Schmidt, U. J. Schmidt, M. Rahlff, E. Schroder, and W. Thust, *Appl. Opt. 11*:1732 (1972).
3. R. Pepperl, *Opt, Acta 24*:413 (1977).
4. I. P. Kaminow and E. H. Turner, *Proc. IEEE 54*:1374 (1966).
5. J. M. Ley, *Electron. Lett. 2*:138 (1966).
6. E. H. Turner, *Appl. Phys. Lett 8*:303 (1966).
7. G. Hepner, *IEEE J. Quantum Electron. QE-8*:169 (1972).
8. S. Sullivan, *Proc. E-O '80 Int. (Brighton)*
9. S. W. Thomas, *Proc. 13th Int. Congr. on High Speed Photogr. and Photonics, Tokyo, 1978,* Vol. 189, SPIE (1978), p. 499.
10. L. N. Magdich, *Opt. Spectrosc.* 248 (1969).
11. J. D. Beasley, *Appl. Opt. 10*:1934 (1971).
12. S. W. Thomas, *Proc. Conf. on High Speed Photogr., Toronto, 1976,* SPIE, Vol. 97, (1976), p. 73.
13. J. M. Ley, T. M. Christmas, and C. G. Wildey, *Proc. IEEE 117*:1057 (1970).
14. B. Cunin, J. A. Miehe, B. Sipp, M. Ya. Schelev, J. N. Serduchenko, and J. Thebault, *Rev. Sci. Instrum. 51*:103 (1980).
15. B. V. Gisin, O. K. Sklyarov, and O. A. Herdochnikov, *Sov. J. Quantum Electron. 5*:248 (1975).
16. C. L. M. Ireland, *Opt. Commun. 27*:459 (1978).
17. C. L. M. Ireland, *Opt. Commun. 30*:99 (1979).
18. C. L. M. Ireland, *Proc. IVth Nat. Quantum Electron. Conf., Edinburgh UK, Sept. 1979,* Wiley, New York, (1980), p. 87.
19. M. Born and E. Wolf, *Principles of Optics*, 5th ed., Pergamon, London (1975).
20. H. Koester, *Electron. Lett. 3*:54 (Feb. 1967).
21. Y. Ninomiya, *IEEE J. Quantum Electron. QE-9*:791 (1973).
22. Y. Ninomiya, *IEEE J. Quantum Electron. QE-10*:358 (1974).
23. I. P. Kaminow and L. W. Stulz, *IEEE J. Quantum Electron. QE-11*:633 (1975).
24. C. S. Tsai and P. Saunier, *Appl. Phys. Lett. 27*:248 (1975).
25. H. Sasaki, *Electron. Lett. 18*:295 (1977).
26. P. Saunier, C. S. Tsai, I. W. Yao, and Le T. Nguyen, *Tech. Digest of Opt. Soc. Am. Meeting on Integrated and Guided Wave Optics, Washington D.C.*, paper TuC2 (1978).
27. J. F. Revelli, *Appl. Opt. 19*:389 (1980).
28. C. H. Bulmer, W. K. Burns, and T. G. Giallorenze, *Appl. Opt. 18*:3282 (1979).
29. F. Auracher, R. Keil, and K. H. Zeitler, *Sixth Europ. Conf. on Opt. Commun., IEEE* (Sept. 1980), pp. 272-275.
30. C. M. Veber and R. P. Kenan, *Sixth Europ. Conf. on Opt. Commun., IEEE*, pp. 124-125.

31. D. Sarid, P. J. Cressmann, and R. L. Holman, *Appl. Phys. Lett.* *33*:514 (1978).
32. V. Ramachandran, *J. Phys. D.* *12*:2223 (1979).

5

Applications for Electro-Optic Deflectors

The electro-optic effect is observed in certain crystals and liquids as a change in refractive index produced by an applied electric field. As a technique for the interaction between electrical signals and light there is no doubt that this involves an attractively simple principle, and in consequence many electro-optic devices have been investigated for the control of light.

The obvious advantage for the electro-optic deflector is that it can be treated as a capacitor, with the speed of operation dependent on the output characteristics of the drive circuitry. In the most advantageous situations nanosecond sweep rates can be obtained with picosecond resolution, and this has led to the construction of prototype streak cameras using bulk multiple-prism or gradient analog deflectors. A commercially available crystal streak camera having 40-psec resolution and a visual display produced by a photodiode array and video processor is shown in Fig. 5.1.

Perhaps the most spectacular application for electro-optic deflectors has been in the area of digital laser displays. Digitally scanned continuous-wave (CW) laser systems are well suited for alphanumeric and graphic high-brightness displays. Real-time large-screen displays have already been used in the presentation of information for military use, and it is suggested that they may eventually become more widely used as a visual aid in traffic control. Only bulk electro-optic digital deflectors have achieved the necessary resolution for the presentation of information in this way. An impressive display indicating the performance achieved by Philips' DLD system is shown in Fig. 5.2

An ever-increasing interest is being directed to the development of beam deflectors for integrated optics, and here nearly all recent work has occurred in the area of surface electro-optic prism and Bragg deflectors. Surface prism and Bragg deflectors are planar devices where a guided beam travels just below a set of strip electrodes. They can achieve a high electric field strength for a relatively low applied voltage and yet have a device capacitance of only a few picofarads. Being quasi-two-dimensional, they can be made using manufacturing techniques developed for integrated optics and can be prepared as an integrated component on a crystal surface.

Devices with moderate resolutions have already been built with base bandwidths approaching 1GHz; consequently they are attractive devices for many fiber-optic communication and integrated-optics applications. These include multiport optic deflection, ultrahigh-data-rate optic multiplexing, and demultiplexing and high-capacity optic

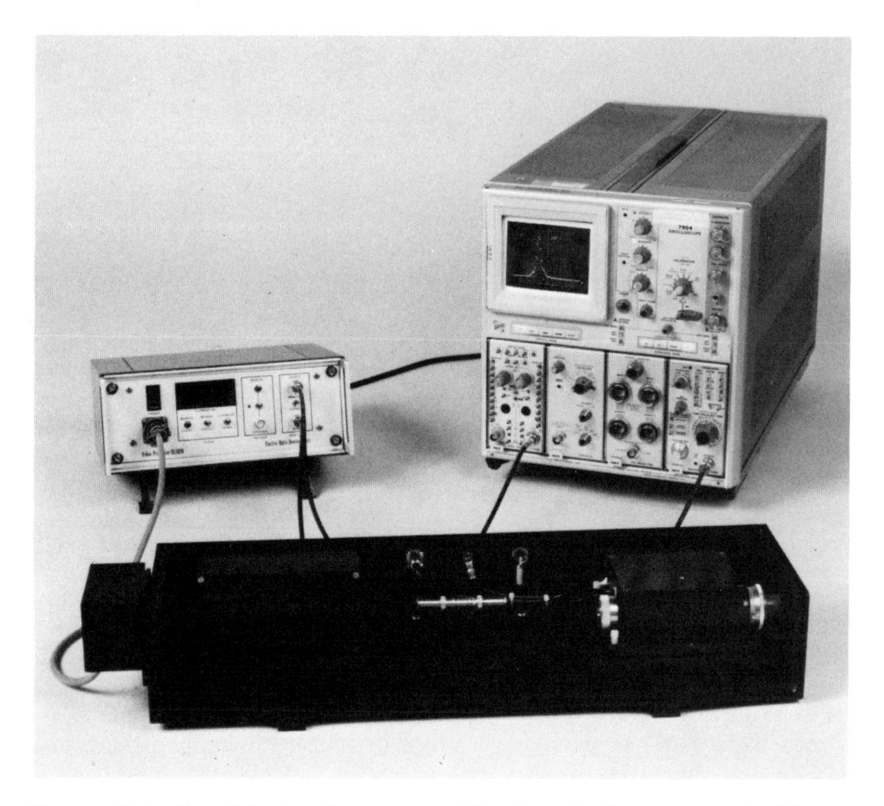

Figure 5.1 Crystal streak camera with photodiode array and video processor. (Published with permission of Electro-Optic Developments Ltd., Basildon, England.)

connecting networks. Surface prism deflectors can also be applied directly with diode array detectors to form analog to digital converters, and they have the potential there to produce a novel range of high-speed devices. In the area of computer development optic memories using holographic or localized storage techniques are also being investigated as potential solutions to the requirement for high-capacity stores of short access time. High-speed deflection of the reference read-in/read-out beam is a fundamental requirement for any optic memory system, and the electro-optic deflector is capable of achieving a suitably short access time. It is considered by those involved in this work that optic memory systems could become commercially available within the next 10 years.

Finally, in the area of laser control, E-O deflectors could be used as intracavity Q spoilers and mode lockers or externally as displacement modulators or pulse shapers. In the latter case the generation

DIGITALE LASERSTRAHLABLENKSTUFE

DIE GRUNDELEMENTE EINER LASERSTRAHLABLENK-STUFE SIND EINE KERR-ZELLE UND EIN KALKSPAT-PRISMA. JE NACH DER POLARISATIONSRICHTUNG DES EINFALLENDEN LICHTES WIRD DER LASERSTRAHL IM DOPPELBRECHENDEN PRISMA ENTWEDER NACH OBEN (SIEHE NEBENSTEHENDES BILD A) ODER NACH UNTEN (SIEHE BILD B) GEBROCHEN. IN BILD A LIEGT DIE POLARISATIONSRICHTUNG DES LASERSTRAHLES PARALLEL ZUR OPTISCHEN ACHSE DES KALKSPATPRISMAS, WAEHREND SIE IN BILD B SENKRECHT DAZU GERICHTET IST. DIE BEIDEN ABLENKRICHTUNGEN SIND DURCH DIE EIGENSCHAFTEN DES DOPPELBRECHENDEN PRISMAS FESTGELEGT. DAVON ABWEICHENDE RICHTUNGEN SIND NICHT MOEGLICH.

BEVOR DER LASERSTRAHL IN DAS KALKSPATPRISMA EINTRITT, DURCHLAEUFT ER EINE KERR-ZELLE. DIESE GESTATTET, DIE POLARISATIONSEBENE DES LICHTES ELEKTRONISCH IN DIE GEWUENSCHTE LAGE ZU STEUERN. DIE KERR-ZELLE BESTEHT AUS ZWEI ELEKTRODEN, ZWISCHEN DENEN SICH EINE ELEKTRO-OPTISCH AKTIVE FLUESSIGKEIT BEFINDET. OHNE SPANNUNG AN DEN ELEKTRODEN (BILD C) DURCHLAEUFT DER LICHTSTRAHL (MIT DER POLARISATIONSRICHTUNG -45°) DIE KERR-ZELLE UNBEEINFLUSST. DURCH ANLEGEN EINER SPANNUNG WIRD DIE FLUESSIGKEIT OPTISCH ANISOTROP. BEI EINER BESTIMMTEN FELDSTAERKE WIRD DANN IN DER FLUESSIGKEIT DIE POLARISATIONSEBENE DES LICHTES UM 90° GEDREHT (BILD D). DER LASERSTRAHL KANN ALSO DURCH ANLEGEN EINER SPANNUNG AN BEIDE ELEKTRODEN ODER DURCH KURZSCHLUSS DES ELEKTRODENPAARES IN DIE EINE ODER DIE ANDERE RICHTUNG ABGELENKT WERDEN.

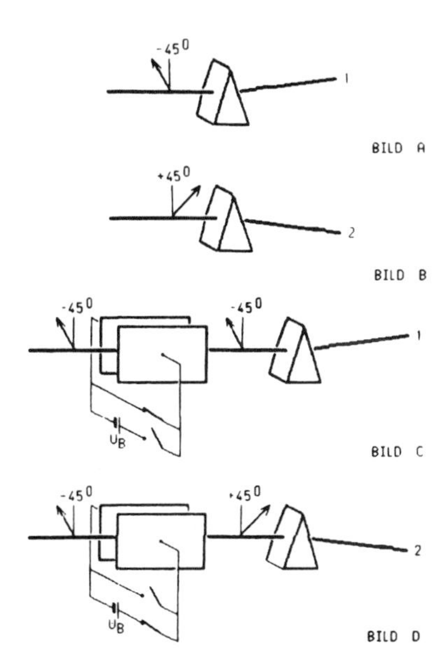

Figure 5.2 Example of the excellent performance achieved by Philips GmBH Hamburg for a real-time scanned laser display employing bulk nitrobenzene E-O digital deflectors. The image comprised 40,000 resolved points. The scanning rate was 12.5 Hz. (See Ref. 1 in Chap. 4 for a description of the system.)

of shaped and short optic pulses in the picosecond regime is of fundamental interest in quantum electronics and of possible value in laser fusion work. For these applications back-to-back synchronized deflectors can be used to realign an optic beam after it has been deflected across an intermediate aperture. The aperture could be a pinhole of Airy disk size to give a very short optic pulse or a slit containing a suitable graded transmission filter to give a desired temporal profile. One such arrangement is shown in Fig. 5.3.

Here a pair of back-to-back synchronized deflectors are used to realign an optical beam after it has been deflected across an intermediate far-field aperture. The aperture can be either a pinhole of approximately Airy disk size to give a very short optical pulse or a slit containing a suitably graded transmission filter to give a desired temporal profile. Using the $LiNbO_3$ deflector described in Chap. 3 and 4, Ireland demonstrated that this arrangement had the potential of yielding pulses shaped on a ~100-psec time scale.

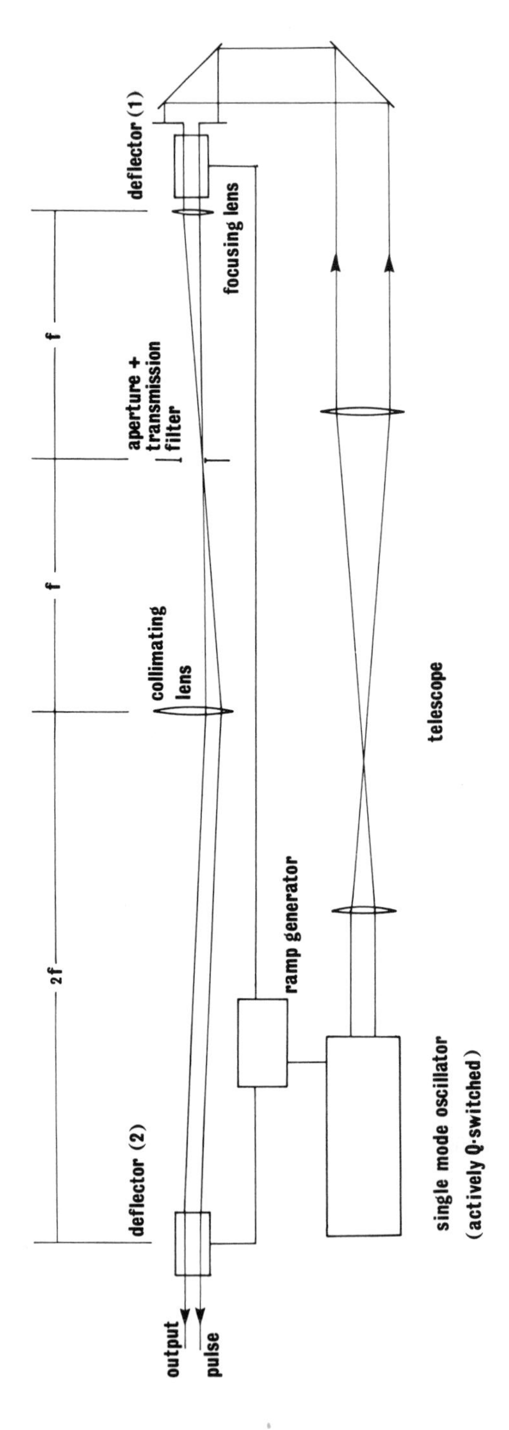

Figure 5.3 Picosecond pulse shaping with E-O deflectors. (From Ref. 18 in Chap. 4.)

There is a common requirement in all the above E-O deflector applications for high-speed optic deflection. In all cases a random access time of less than 1 μsec is necessary, and in certain cases less than 10 nsec is required. It is in this area that the E-O deflector has unique advantages. Apart from the digital Kerr cell deflector, E-O devices have not yet achieved linear resolutions of above 1000 spot positions, and it would seem that at least for the immediate future their use will be confined to moderate-resolution high-speed or very high-speed applications.

LIST OF SYMBOLS

B	Kerr constant for quadratic E-O liquid
$\underline{D}$	Electric displacement vector
d_{ij}	Linear piezoelectric coefficient
$\underline{E}$	Electric field vector
$g_{ijk\ell}$	Induced polarization coefficient
$h_{ink\ell}$	Quadratic E-O coefficient
M_R	Figure of merit for spatial resolution of deflector
M_τ	Figure of merit for temporal resolution of deflector
N_R	Number of resolvable optic spots
$\underline{P}$	Polarization vector
P_r	Reactive drive power
p_{cj}	Strain-optic coefficient
R	Linear capacity of digital light deflector (DLD)
R_e	Square resistance of electrodes
r_{ij}	Linear E-O coefficient
S/N	Signal to noise ratio
T_c	Transition temperature
$\tan\delta$	Loss tangent
V_{app}	Applied voltage
ε_0	Permittivity of free space susceptibility
θ_R	Diffraction-limited optic beam divergence
ψ	Plane angle of harmonic wave

part II
ACOUSTO-OPTIC DEFLECTORS

MILTON GOTTLIEB / Westinghouse Research and Development Center, Pittsburgh, Pennsylvania

Despite the relative simplicity of the physical effects involved in acousto-optic interactions, no entirely satisfactory microscopic theory of the photoelastic effect has been widely accepted. We shall, therefore, present a simple phenomenological theory that will be entirely adequate to understand all the observed behavior in both isotropic and anisotropic materials. This will also provide the analytic framework for designing and optimizing a variety of acousto-optic devices and systems, several of which are now commonly available from a number of commercial sources. These commercially available devices rely upon a few outstanding materials and offer a limited range of operating characteristics. Design principles will be presented so that these outstanding materials can be used for acousto-optic systems with greatly extended high-performance characteristics. In addition, other efficient materials, which may possess certain unique properties making them desirable in specialized systems, will also be described. Such materials will be required, for example, for systems operating at ultraviolet or infrared wavelengths inaccessible to crystals such as $PbMoO_4$, TeO_2, or $LiNbO_3$.

Ultrasonic techniques for the VHF range are crucial to the fabrication of acousto-optic devices, and a large body of literature exists devoted specifically to this subject. The chapter in this volume on acoustic techniques deals only very generally with the fundamentals of acoustic design but in greater depth as applied specifically to acousto-optic devices. Perhaps the best example of this is the discussion devoted to the techniques to accomplish acoustic beam steering for Bragg angle matching. These concepts take their inspiration from the theory of radar and sonar electronic beam steering but are adapted to the transducer structures and propagation characteristics of acousto-optic devices.

Finally, the relatively new and still speculative field of integrated optics, or guided wave optics, will be described in the final chapter. Again, it is not possible here to discuss this entire subject in any detail; only the most basic phenomena will be described as required to understand acousto-optic interactions in thin-film waveguides. Such interactions are directly analogous to those in bulk systems but open a whole regime of operation for scanning and deflection with greatly miniaturized components and very low drive powers. Thin-film waveguides have the potential to revolutionize optics in much the same way that electronics was revolutionized by thin-film integration. Acousto-optics will undoubtedly play an important role in this new revolution.

6

Acousto-Optic Interactions

6.1 THE PHOTOELASTIC EFFECT

The underlying mechanism of all acousto-optic interactions is very simply the change induced in the refractive index of an optic medium due to the presence of an acoustic wave. An acoustic wave is a traveling pressure disturbance which produces regions of compression and rarefaction in the material. The refractive index is related to the density, for the case of an ideal gas, by the Lorentz-Lorenz relation

$$\frac{n^2 - 1}{n^2 + 2} = \text{constant} \times \rho \tag{6.1}$$

where n is refractive index and ρ is density. In fact, this relation is adhered to remarkably well for most simple solid materials as well. The elasto-optic coefficient is obtained directly by differentiation of (6.1):

$$\rho \frac{\partial n}{\partial \rho} = \frac{(n^2 - 1)(n^2 + 2)}{6n} \tag{6.2}$$

where it is understood that the derivative is taken under isentropic conditions. This is generally the case for ultrasonic waves in which the flow of energy by thermal conduction is slow compared with the rate at which density changes within a volume smaller than an acoustic wavelength. The fundamental quantity given by Eq. (6.2), also known as the photoelastic constant p, can be easily related to the pressure applied, with the result

$$P = \frac{1}{\beta} \frac{\partial n}{\partial P} \tag{6.3}$$

where P is the applied pressure and β is the compressibility of the material. The photoelastic constant of an ideal material with refractive index of 1.5 is 0.59. It will be seen later that the photoelastic constants of a wide variety of materials lie in the range from about 0.1 to 0.6, so that this simple theory gives a reasonably good approximation to measured values.

The relation in Eq. (6.3) follows from the usual definition of the photoelastic constant:

$$\Delta\left(\frac{1}{\varepsilon}\right) = \Delta\left(\frac{1}{n^2}\right) = pe \tag{6.4}$$

where ε is the dielectric constant ($\varepsilon = n^2$) and e is the strain amplitude produced by the acoustic wave. From Eq. (6.4) it is easily seen that the change in refractive index Δn produced by the strain is

$$\Delta n = -\frac{1}{2} n^3 pe \tag{6.5}$$

where e is of the form $e_0 \exp(i\Omega t)$ for an acoustic wave of frequency Ω. The magnitude of the changes in refractive index that are typical for acousto-optic devices are not large. Strain amplitudes lie in the range 10^{-8} to 10^{-5}, so that using the above expressions for Δn and p gives for Δn about 10^{-8} to 10^{-5} (for $n = 1.5$). It may be somewhat surprising, then, that devices based on such a small change in refractive index are capable of generating large effects, but it will be seen that this comes about because these devices are configured in a way that can produce large phase changes at optical wavelengths.

The relation defining the photoelastic interaction has been written in Eq. (6.5) as a scalar relation in which the photoelastic constant is independent of the directional properties of the material. In fact, even for an isotropic material such as glass, longitudinal acoustic waves and transverse (shear) acoustic waves cause the photoelastic interaction to assume different parameters. A complete description of the interaction, particularly for anisotropic materials, requires a tensor relation between the dielectric properties, the elastic strain, and the photoelastic coefficient. This may be represented by the tensor equation

$$\Delta\left(\frac{1}{n^2}\right)_{ij} = \sum_{k\ell} P_{ijk\ell} e_{k\ell} \tag{6.6}$$

where $(1/n^2)_{ij}$ is a component of the optic index ellipsoid, $e_{k\ell}$ are the Cartesian strain components, and $p_{ijk\ell}$ are the components of the photoelastic tensor. The crystal symmetry of any particular material determines which of the components of the photoelastic tensor may be nonzero and also which components are related to others. This may be useful in determining whether or not some crystal, based only on its symmetry, may even be considered for certain applications.

6.2 DIFFRACTION BY ACOUSTIC WAVES

The most useful photoelastic effect is the ability of acoustic waves to diffract a light beam. There are several ways to understand how diffraction comes about; the acoustic wave may be thought of as a diffraction grating made up of periodic changes in optic phase, rather than transparency, and moving at sonic velocity, rather than being stationary. Thus, it is possible to analyze the diffraction as resulting from a moving phase grating. Alternatively, the light and sound may be thought of as particles, photons and phonons, undergoing collisions in which energy and momentum are conserved. Either of these descriptions may be used to obtain all the important diffraction effects, but some are more easily understood on the basis of one or the other. It will be useful, then, to outline both of these approaches.

To examine the simplest case of plane acoustic waves interacting with plane light waves, consider the diagram shown in Fig. 6.1. Suppose the light wave, of frequency ω and wavelength λ, is incident from the left into a delay line with an acoustic wave of frequency Ω and wavelength Λ. If the refractive index of the delay medium is $n + \Delta n$ in the presence of the acoustic wave, the phase of the optic wave will be changed by an amount

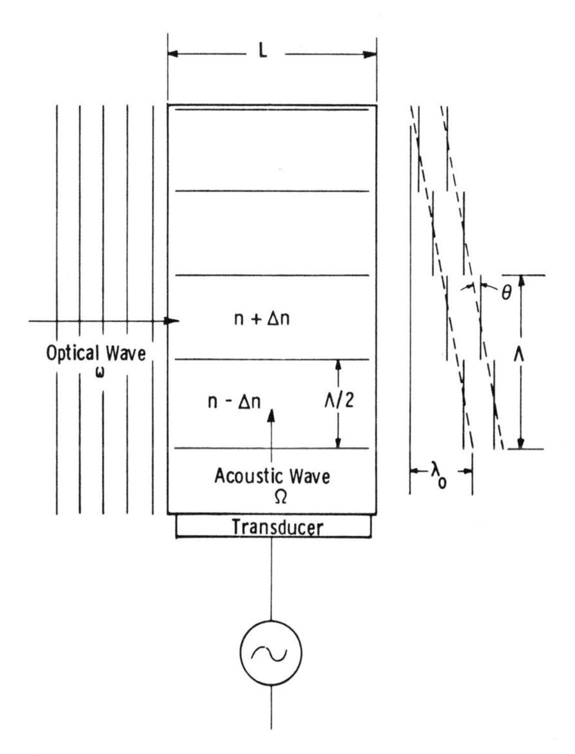

Figure 6.1 Tilting of optical wavefronts caused by upward-traveling acoustic wave.

$$\Delta\phi = 2\pi \frac{L}{\lambda} \Delta n \tag{6.7}$$

if the width of the delay line is L. Some typical values of $\Delta\phi$ can be obtained by assuming L = 2.5 cm and λ = 0.5 μm, with Δn reaching a peak value of 10^{-5}. This yields a phase change of π rad, which is, of course, quite large. It is large because L/λ, the number of optic wavelengths, is 50,000, so that a very small Δn can still produce a sizable $\Delta\phi$. If the electric field incident on the delay line is represented by

$$E = E_0 e^{i\omega t} \tag{6.8}$$

then the field of the phase-modulated emerging light will be

$$E = E_0 e^{i(\omega t + \Delta\phi)} = e^{i\omega t} e^{i2\pi(L/\lambda)(a_0 \sin \Omega t)} \tag{6.9}$$

We shall not give here a detailed derivation of the resulting temporal and spatial distribution of the light field, but we can use intuition and analogy with radio-wave modulation to arrive at the resultant fields. It is a well-known RF engineering concept that the spectrum of a phase-modulated carrier of frequency ω consists of components separated by multiples of the modulation frequency Ω, as shown in Fig. 6.2. There is a multiplicity of sidebands about the carrier frequency such

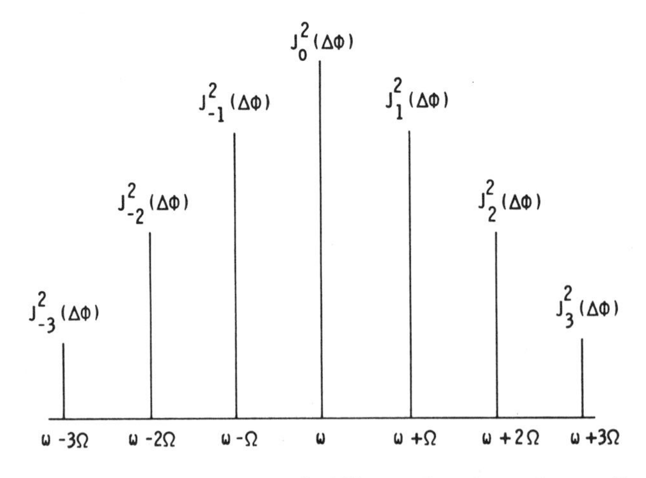

Figure 6.2 Intensity of diffracted orders due to Raman-Nath interaction.

that the frequency of the nth sideband is $\omega + n\Omega$, where n is both positive and negative. The amplitudes of each of the sidebands are proportional to the Bessel function of order equal to the sideband number and whose argument is the modulation index $\Delta\phi$. Although not shown by Fig. 6.2, note that the odd-numbered negative orders are 180° out of phase with the others. The light emerging from the delay line is composed of a number of light waves whose frequencies have been shifted by $n\Omega$ from the frequency ω of the incident light. The relative amplitudes will be determined by the peak change in the refractive index.

To understand the diffraction of the light by the acoustic wave, consider the optical wavefronts shown in Fig. 6.1. Since the velocity of light is about five orders of magnitude greater than the velocity of sound, it is a good approximation to assume that the acoustic wave is stationary in the time it takes the optic wave to traverse the delay line. Suppose that during this instant the half-wavelength region labeled $n + \Delta n$ is under compression and $n - \Delta n$ under rarefaction. Then the part of the optic plane wave passing through the compression will be slowed (relative to the undisturbed material of index n), while the part passing through the rarefaction will be speeded up. In this rough picture, the emerging wavefront will be *corrugated*, so that if the corrugations are joined by a continuous plane, its direction is tilted relative to that of the incident light wavefronts. Since the optical phase changes by 2π for each acoustic wavelength Λ along the acoustic beam direction, the tilt angle will be given by $\theta \simeq \lambda/\Lambda$. The direction normal to the tilted plane is the direction of optic power flow and represents the diffracted light beam. Note that the corrugated wavefront could just as well have been connected by a tilted wavefront at an angle given by $\theta \simeq -(\lambda/\Lambda)$; this corresponds to the first negative order and the other to the first positive order. A similar argument for the higher sidebands leads to a complete set of deflected beams, as shown in Fig. 6.3, where the angular deflection corresponding to the nth order is given by $\theta_n \simeq \pm n(\lambda/\Lambda)$, and the frequency of the light deflected into the nth order is $\omega \pm n\Omega$. The intensity of the carrier wave, or zeroth order, will be zero when the modulation index $\Delta\phi$ is equal to 2.4. The generally important first order will be a maximum for $\Delta\phi = 1.8$, decreasing for higher modulation. These phenomena were described by Debye and Sears [1] and are often referred to as Debye-Sears diffraction. Similar observations were published almost simultaneously by Lucas and Biquard [2]. An extensive theoretical analysis of the effect was given by Raman and Nath [3], and so it is alternatively referred to as Raman-Nath diffraction. A distinctive feature of this type of diffraction is that it is limited to low acoustic frequencies (or relatively long wavelengths). The origin of this limitation lies in the diffraction spreading of the light beam as it traverses apertures formed by the columns of compression and rarefaction in the acoustic beam. If the length of the acoustic beam along the light

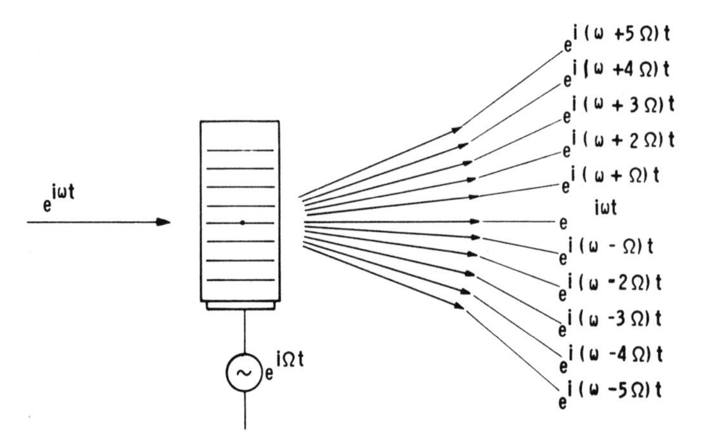

Figure 6.3 Raman-Nath diffraction of light into multiple orders.

propagation direction is large enough, the diffraction spread of the light between adjacent compression and rarefaction regions will overlap, so that there is some maximum length of interaction region beyond which the Debye-Sears effect smears out. To estimate this maximum length, suppose the compression and rarefaction apertures are one-half an acoustic wavelength $\Lambda/2$, so that the angular diffraction spread of the light is $\delta\phi \simeq 2\lambda/\Lambda$. Then ℓ_{max} can be defined as that interaction length for which the aperture diffraction spreads the light by one-half an acoustic wavelength:

$$\ell_{max}\,\delta\phi = \frac{\Lambda}{2} \tag{6.10}$$

or

$$\ell_{max} = \frac{\Lambda^2}{4\lambda} \tag{6.11}$$

This can be expressed as

$$Q \equiv \frac{4\ell\lambda}{\Lambda^2} < 1$$

where the quantity Q, known as the Raman-Nath parameter,† relates ℓ, λ, and Λ for which the *thin grating* approximation is valid. For

†Often referred to as the Klein-Cook parameter.

typical values of $\ell = 1$ cm and $\lambda = 6.33 \times 10^{-5}$ cm, $Q = 1$ for $\Lambda =$ 0.0159 cm, which corresponds to a frequency of 31.4 MHz for a material whose acoustic velocity is 5×10^5 cm/sec.

For values of the interaction length $\ell > \ell_{max}$ for which the thin grating approximation no longer holds, a different regime of operation takes effect. If the incident light beam is normal to the sound beam propagation direction, the higher diffraction orders interfere destructively beyond ℓ_{max}, eventually completely wiping out the diffraction pattern. For constructive interference to take place, the angle of incidence must be tilted with respect to the acoustic beam direction. To better understand what conditions must be satisfied for this, it is easier to think of the light and sound waves as colliding photons and phonons. In this description, the light and sound take on the attributes of particles, and the dynamics of their collisions are governed by the laws of conservation of energy and conservation of momentum. The magnitudes of the momenta of the light and sound waves are given by the well-known expressions

$$|\underline{k}| = \frac{\omega n}{c} = \frac{2\pi n}{\lambda_0} \tag{6.12}$$

and

$$|\underline{K}| = \frac{\Omega}{v} = \frac{2\pi}{\Lambda} \tag{6.13}$$

respectively. In the latter equation, v is the velocity of sound in the delay medium, $v = 2\pi\Omega\Lambda$. Conservation of momentum is expressed by the vector relation

$$\underline{k}_i + \underline{K} = \underline{k}_d \tag{6.14}$$

the diagram for which is shown in Fig. 6.4a, where $\underline{k}_i$ and $\underline{k}_d$ represent the momentum of the incident photon and the diffracted photon, respectively. The process may be thought of as one in which the acoustic phonon is absorbed by the incident photon to form the diffracted photon. Thus, conservation of energy requires that

$$h\omega_0 = h\omega_i + h\Omega \tag{6.15}$$

or

$$\omega_d = \omega_i + \Omega$$

in which h is Planck's constant.

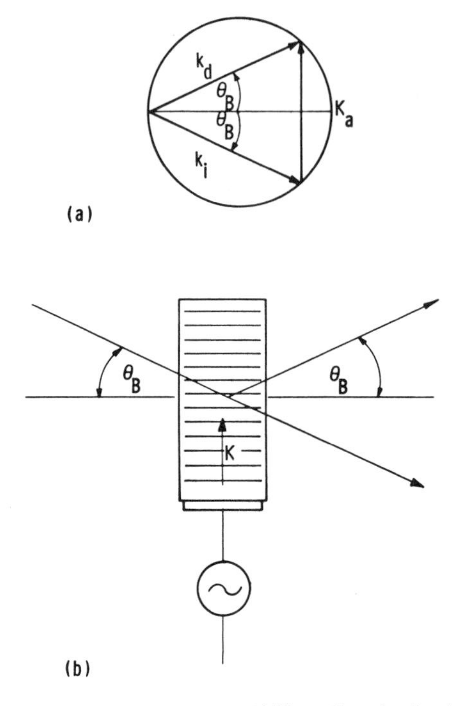

Figure 6.4 Bragg diffraction in isotropic medium.

Since ω_i lies in the optic frequency range and Ω will typically lie in the RF or microwave range, $\Omega >> \omega_i$ so that $\omega_d \cong \omega_i$. This results in the magnitudes of $\underline{k}_i$ and $\underline{k}_d$ being almost equal, so that the momentum triangle of Fig. 6.4a is isosceles, and the angle of incidence (with respect to the normal to $\underline{K}$) is equal to the angle of diffraction. This angle is easily obtained from Fig. 6.4a as

$$\sin \Theta_B = \frac{1}{2}\frac{K}{k} = \frac{1}{2}\frac{\lambda}{\Lambda} \tag{6.16}$$

It is called the Bragg angle because of its similarity to the angle of diffraction of X-rays from the regularly spaced planes of atoms in crystals. The configuration of these vectors in relation to the delay line is shown in Fig. 6.4b. For diffraction to take place, the light must be incident at the angle Θ_B, and the diffracted beam will appear only at this same angle. In contrast to the Debye-Sears regime, there are no higher-order diffracted beams. In the full mathematical treatment of the Bragg limit ($Q = 4\ell\lambda/\Lambda^2 >> 1$) light energy may appear at the higher orders, but the probability of its doing so is extremely small, so that the intensity at higher orders is essentially zero. The diagrams in Fig. 6.4 show the interaction in which the diffracted photon is higher in energy than that of the incident photon, but the reverse can also take place. If the sense of the vector $\underline{K}$ is reversed with respect to $\underline{k}_i$, then $\omega_d = \omega_i - \Omega$ and the diffracted negative first order results.

It is important to understand that the Debye-Sears effect and Bragg diffraction are not different phenomena but are the limits of the same mechanism. The Raman-Nath parameter Q determines which is the appropriate limit for a given set of values λ, Λ, and ℓ. Quite commonly in practice these values will be chosen such that neither limit applies, and $Q \approx 1$. In this case, the mathematical treatment is quite complex, and experimentally it is found that one of the two first-order diffracted beams may be favored but that higher orders will be present.

6.3 DIFFRACTION EFFICIENCY

Having obtained the angular behavior of light diffracted by acoustic waves, the next most important characteristic is the intensity of the diffracted beam. Again, the full mathematical treatment is beyond the scope of this book, but a very good intuitive calculation leads to results that are useful. Referring to the spectrum of a phase-modulated wave shown in Fig. 6.2, we can see that the ratio of the intensity in the first order to that in the zero order is

$$\frac{I_1}{I_0} = \left[\frac{J_1(\Delta\phi)}{J_0(\Delta\phi)}\right]^2 \tag{6.17}$$

We shall now show in detail how this result comes about for acousto-optically diffracted light. The acoustic power flow is given by

$$P = \frac{1}{2} cve^2 \tag{6.18}$$

where c is the elastic stiffness constant. The elastic stiffness constant is related to the bulk modulus β and the density and acoustic velocity through the well-known expression

$$c = \frac{1}{\beta} = \rho v^2 \tag{6.19}$$

Thus, the acoustic power density is

$$P_A = \frac{1}{2} \rho v^3 e^3 \tag{6.20}$$

We can express the phase modulation depth in terms of the acoustic power density, using Eqs. (6.5) for Δn and (6.7) for $\Delta\phi$, with the result

$$\Delta\phi = 2\pi \frac{L}{\lambda}\Delta n = -\pi \frac{L}{\lambda} n^3 p \left(\frac{2P_A}{\rho v^3}\right)^{1/2} \tag{6.21}$$

For small modulation index, the zero-order and first-order Bessel functions can be approximated by

$$\begin{aligned} J_0(\Delta\phi) &\approx \cos(\Delta\phi) \approx 1 - \Delta\phi \\ J_1(\Delta\phi) &\approx \sin(\Delta\phi) \approx \Delta\phi \end{aligned} \tag{6.22}$$

so that the small signal approximation to the diffracted light is, from Eq. (6.17),

$$\frac{I_1}{I_0} \approx (\Delta\phi)^2 = \frac{\pi^2}{2}\left(\frac{L}{\lambda}\right)^2 \left(\frac{n^6 p^2}{\rho v^3}\right) P_A \tag{6.23}$$

This efficiency may be expressed in terms of the total acoustic power P:

$$P = P_A(LH) \tag{6.24}$$

where H is the height of the transducer and

$$\frac{I_1}{I_0} = \frac{\pi^2}{2}\frac{L}{H}\left(\frac{n^6 p^2}{\rho v^3}\right)\frac{P}{\lambda^2} \tag{6.25}$$

The quantity in parentheses depends only on the intrinsic properties of the delay line material, while the other parameters depend on external factors. It is therefore defined as the figure of merit of the material,

$$M_2 \equiv \left(\frac{n^6 p^2}{\rho v^3}\right) \tag{6.26}$$

from which it can be seen that, in general, the most important factors leading to high acousto-optic efficiency will be a high refractive index and a low acoustic velocity. This does not guarantee a large figure of merit, since the photoelastic constant may be very small, or even zero.

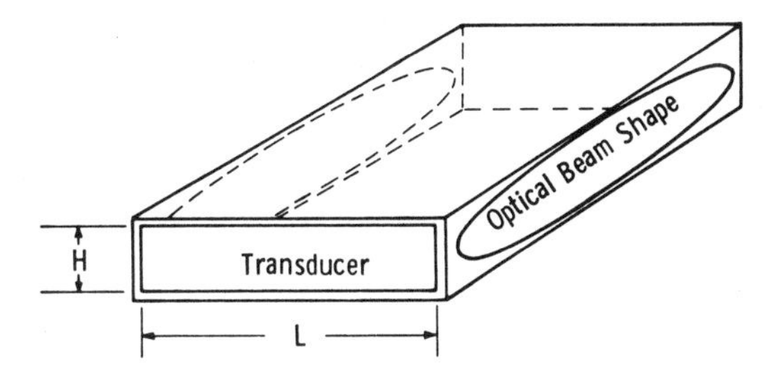

Figure 6.5 Transducer and optical beam shapes for optimization of acousto-optic diffraction.

The other factors in Eq. (6.25) have the following effect on the diffraction efficiency. The efficiency decreases quadratically with increasing wavelength, so that the power requirements for operation in the IR may be hundreds of times that required for the visible. For high efficiency, it will be desirable to have a large aspect ratio L/H, leading to a configuration as shown in Fig. 6.5. It is difficult to make conventional bulk devices with H much less than 1 mm, so that aspect ratios of about 50 can be achieved. Much higher aspect ratios can be reached in guided optic wave devices. A more exact calculation of the diffraction efficiency in the Bragg regime [4] yields the result

$$\frac{I_1}{I_0} = \sin^2\left(\frac{\pi^2}{2}\frac{L}{H}M_2\frac{P}{\lambda^2}\right)^{1/2} \tag{6.27}$$

For low signal levels Eq. (6.27) reduces to the same expression as in (6.25). To obtain an order of magnitude for the power requirements of an acousto-optic deflector, let us assume a material with $n = 1.5$, $\rho = 3$, $v = 5 \times 10^5$ cm/sec, and a photoelastic constant calculated from the Lorentz-Lorenz expression $p \simeq 0.6$, so that $M_2 \approx 1.1 \times 10^{-17}$ sec^3/g. If the remaining parameters are $L = 1$ cm and $\lambda = 0.6$ μm, then by assuming a maximum acoustic power density for CW operation of 1 W/cm^2 (10^7 erg/cm^2 sec), the maximum obtainable efficiency is 15%. We shall see in the later chapters, however, that materials and designs are available which are capable of realizing higher efficiencies with lower power levels.

6.4 ANISOTROPIC DIFFRACTION

The theory of diffraction of light thus far presented has assumed that the optic medium is isotropic, or at least that it is not birefringent. A number of important acousto-optic devices make use of the properties of birefringent materials, so a brief description of the important characteristics of anisotropic diffraction will be given here. The essential difference from diffraction in isotropic media is that the momentum of the light

$$k = \frac{2\pi}{\lambda} = \frac{2\pi n}{\lambda_0} \tag{6.28}$$

will, in general, be different for different light polarization directions. Thus, the vector diagram representing conservation of momentum will no longer be the simple isosceles triangle of Fig. 6.4.

To understand the effect of anisotropy on diffraction, it is necessary to mention another phenomenon which occurs when light interacts with shear acoustic waves, that is, waves in which the displacement of matter is perpendicular to the direction of propagation of the acoustic wave. A shear acoustic will cause the direction of polarization of the diffracted light to be rotated by 90°. The underlying reason for this is that the shear disturbance induces a birefringence which acts upon the incident light as a birefringent plate, that is, rotates the plane of polarization. This phenomenon occurs in isotropic materials as well as in anisotropic materials. However, in isotropic materials the magnitude of the momentum vector $k = 2\pi n/\lambda_0$ will be the same for both polarizations, so there is no effect upon the diffraction relations. Suppose, instead, that the interaction occurs in a birefringent crystal in a plane perpendicular to the optic axis. Let us choose the example as shown in the index surfaces in Fig. 6.6 in which the incident light is an extraordinary ray. For this example,

$$k_i = \frac{2\pi n_e}{\lambda_0} \quad \text{and} \quad k_d = \frac{2\pi n_o}{\lambda_0} \tag{6.29}$$

and the angle of incidence θ_i and diffraction θ_d are in general not equal. The theory of anisotropic diffraction was developed by Dixon [5] in whose work the expressions for the anisotropic Bragg angles were derived as

$$\sin\theta_i = \frac{1}{2n_i}\frac{\lambda_0 f}{v}\left[1 + \left(\frac{v}{\lambda_0 f}\right)^2 (n_i^2 - n_d^2)\right] \tag{6.30}$$

$$\sin\theta_d = \frac{1}{2n_d}\frac{\lambda_0 f}{v}\left[1 - \left(\frac{v}{\lambda_0 f}\right)^2 (n_i^2 - n_d^2)\right] \tag{6.31}$$

where n_i and n_d are the refractive indices corresponding to the incident and the diffracted light polarizations and f is the acoustic frequency:

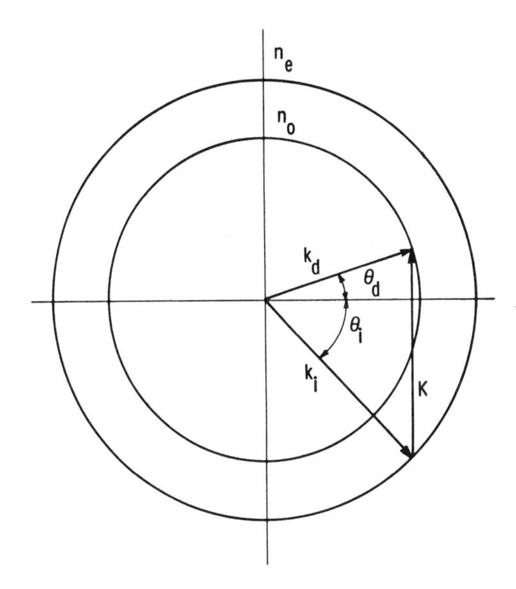

Figure 6.6 Vector diagram for diffraction in birefringent medium.

$$f = \frac{v}{\Lambda} \tag{6.32}$$

These angles are plotted in Fig. 6.7 for a particular set of parameters: $n_i = 2.0$, $n_d = 1.9$, $v = 10^5$ cm/sec, and $\lambda_0 = 1$ μm. These curves, the

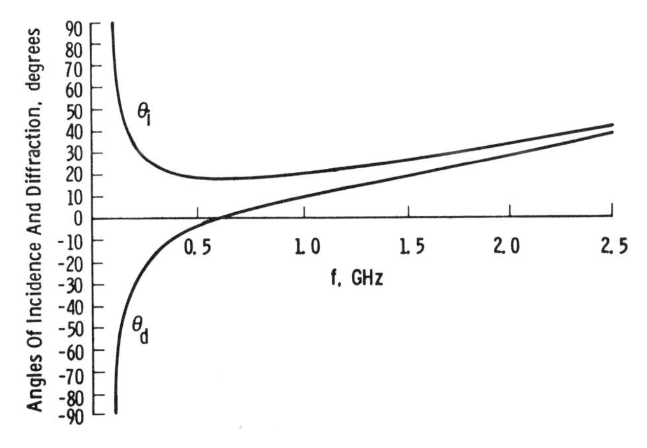

Figure 6.7 Angles of incidence and diffraction for anisotropic medium in which $n_i = 2$, $n_d = 1.9$, $v = 10^5$ cm/sec, and $\lambda = 1$ μm.

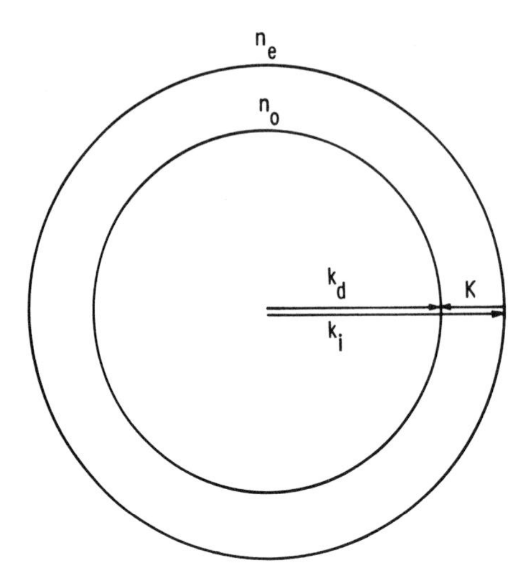

Figure 6.8 Vector diagram for collinear diffraction in birefringent medium.

general shapes of which are similar for all birefringent crystals, have a number of interesting characteristics which are useful for several types of acousto-optic devices. The minimum frequency for which an interaction may take place corresponds to $\theta_i = 90°$ and $\theta_d = -90°$, for which all three vectors will be collinear, as shown in Fig. 6.8. It is easily shown that for this case, since the vector equation for conservation of momentum can be written as a scalar equation,

$$|\underline{k}_i| + |\underline{K}| = |\underline{k}_d| \tag{6.33}$$

the frequency for which collinear diffraction takes place is

$$f = \frac{v(n_i - n_d)}{\lambda_0} \tag{6.34}$$

Such collinear phase matching has been used as the basis of an important device, the electronically tunable acousto-optic filter [6]. Note that if the incident light had been chosen as ordinary rather than extraordinary polarized, the sense of the acoustic vector $\underline{K}$ would be reversed. In fact, the roles of the two curves in Fig. 6.7 would be reversed by interchanging n_i and n_d.

Another interesting region of anisotropic diffraction occurs at the minimum value in the curve representing θ_i at which frequency $\theta_d = 0$. This frequency is obtained by setting the quantity in brackets in Eq. (6.31) equal to zero:

$$f = \frac{v}{\lambda_0} \sqrt{n_i^2 - n_d^2} \tag{6.35}$$

For the chosen parameters of Fig. 6.7, this frequency is 624 MHz. The significance of this point is that the angle of incidence of a scanned beam is relatively insensitive to change over a very broad range of frequencies. For the curves shown in Fig. 6.7, the minimum θ_i is 18.25° and increases to only 20° for a range of frequencies from 400 MHz to 1 GHz, a fractional bandwidth $\Delta f/f_0$ of 85%. As will be seen, it is very difficult to achieve an interaction bandwidth this large by any other method.

The description of the interaction of light with sound we have given above is perhaps the simplest in terms of giving an intuitive understanding of the phenomena. Other descriptions, with totally different mathematical formalisms, have been carried out, and these lead to many details and subtleties in the behavior of acousto-optic systems that are beyond the scope of this book. Exact calculations have been carried out to extend the range of validity [7] from the limits allowed by the Raman-Nath theory [8] and this has been experimentally investigated [9]. Other studies have also been carried out to give accurate numerical results for the intensity distribution of light in the various diffraction orders [10]. The diffraction process has been reviewed and analyzed by Klein and Cook [11] using a coupled mode formulation, and there is continuing recent interest in refining the plane wave scattering theory to give explicit results for intermediate cases [12,13]. Finally, the acousto-optic interaction can be viewed as a parametric process in which the incident optic wave mixes with the acoustic wave to generate polarization waves at sum and difference frequencies, leading to new optical frequencies; this approach has been reviewed by Chang [14].

6.5 RESOLUTION AND BANDWIDTH CONSIDERATIONS

Resolution, bandwidth, and speed are the important characteristics of acousto-optic scanners, shared by all types of scanning devices. Depending on the application, only one or all of these characteristics may have to be optimized; in this section, we shall examine which acousto-optic design parameters are involved in the determination of resolution, bandwidth, and speed. Consider an acousto-optic scanner with a collimated incident beam of width D diffracted to an angle θ_0 at the center of its bandwidth Δf. If the diffracted beam is focused onto a plane by a lens, or lens combination, at the delay line, the diffraction spread of the optical beam will be

$$\delta x = F\ \delta\phi \simeq F\frac{\lambda}{D} \tag{6.36}$$

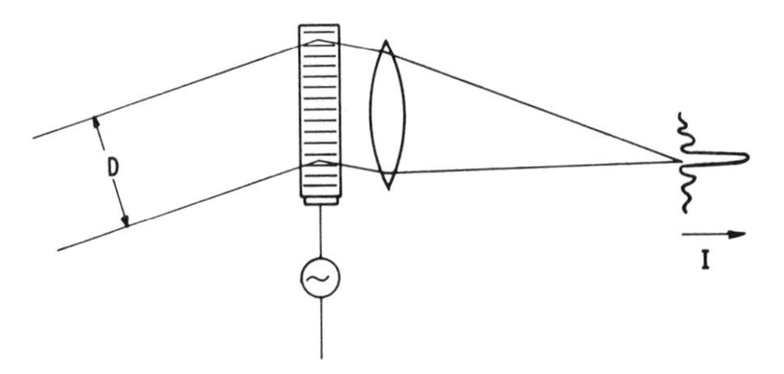

Figure 6.9 Distribution of light intensity due to diffraction by acoustic field.

where F is the focal length of the lens. The light intensity will be distributed in the focal plane as illustrated in Fig. 6.9. As an example, for diffraction-limited optics, the spot size for a 25-mm-wide light beam of wavelength 0.633 μm at a distance of 30 cm from the delay line is 7.6 μm. There are, however, aberrations which prevent this from being fully realized, as will be discussed later. The number of resolvable spots will be the angular scan range divided by the angular diffraction spread:

$$N = \frac{\Delta\theta}{\delta\phi} \tag{6.37}$$

where $\Delta\theta$ is the range of the angular scan. Differentiating the Bragg angle formula yields

$$\Delta\theta = \frac{\lambda}{v \cos\theta_0} \Delta f \tag{6.38}$$

and

$$N = \Delta f \left(\frac{D}{v \cos\theta_0} \right) = \Delta f \tau \tag{6.39}$$

where τ is the time that it takes the acoustic wave to cross the optical aperture. The resulting expression is the *time-bandwidth* product of the delay line, a concept applied to a variety of electronic devices as a measure of information handling capacity. The time-bandwidth product of an acousto-optic delay line is equivalent to the number of bits of information which may be simultaneously processed by the system. To maximize the number of resolution elements, it is desirable to have as large a bandwidth as possible (i.e., large frequency range) and also as large a delay time as possible. There are two factors limiting the bandwidth of an acousto-optic device: the bandwidth of the transducer structure, which will be discussed later, and acoustic absorption in the delay medium. The acoustic absorption increases with increasing frequency. For high-purity single crystals the increase generally goes with the square of the frequency. For glassy materials, on the other hand, the attenuation will increase more slowly, often approaching a linear function. The maximum frequency is generally taken as that for which the attenuation of the acoustic wave across the optical aperture is equal to 3 dB. A reasonable approximation of the maximum attainable bandwidth is $\Delta f = 0.7 f_{max}$, so that we may derive some relationships for the maximum number of resolution elements. For a material with a quadratic dependence of attenuation on frequency,

$$\alpha(f) = \Gamma f^2 \tag{6.40}$$

and the maximum aperture for 3-dB loss is

$$D = \frac{3}{\Gamma f^2} \tag{6.41}$$

Using these results, the maximum number of resolution elements is

$$N_{max} \simeq \sqrt{\frac{1.5D}{v^2 \Gamma}} \tag{6.42}$$

from which it can be seen that in principle it is always advantageous to make the delay line as long as possible. In practice, the aperture will be limited by the largest crystals that can be prepared, or ultimately by the size of the optical system. For a glassy material for which the attenuation increases linearly with frequency,

$$\alpha(f) = \Gamma' f \tag{6.43}$$

and the maximum number of resolvable spots will be

$$N_{max} \simeq \frac{2}{\Gamma' v} \tag{6.44}$$

which is independent of the size of the aperture, being determined only by the material attenuation constant and the acoustic velocity. In the next section we shall review material considerations in some detail and see what the performance limits are of currently available acousto-optic materials. As a numerical example, however, the highest-quality fused quartz has an attenuation of about 3 dB/cm at 500 MHz and an acoustic velocity of 5.96×10^5 cm/sec (for longitudinal waves), leading to $N_{max} = 560$.

6.6 INTERACTION BANDWIDTH

The number of resolution elements will be determined by the frequency bandwidth of the transducer and delay line, but a number of other bandwidth considerations are also of importance for the operation of a scanning system. While a large value of τ leads to a large value of N, the speed of the device is just equal to $1/\tau$. That is, the position of a spot cannot be changed randomly in a time less than τ. If the acoustic cell is being used to temporally modulate the light as well as to scan, then obviously the modulation bandwidth will similarly be limited by the travel time of the acoustic wave across the optical aperture. To increase the modulation bandwidth, the light beam must be focused to a small width w in the acoustic field. The 3-dB modulation bandwidth is approximately

$$\Delta f = \frac{0.75}{\tau} = \frac{0.75v}{w} \tag{6.45}$$

and the diffraction-limited beam waist (the $1/e^2$ power points) of a Gaussian beam is

$$w_0 = \frac{2\lambda_0 F}{\pi D} \tag{6.46}$$

where D is the incident beam diameter and F is the focal length of the lens. With this value of beam waist, the maximum modulation bandwidth is

$$\Delta f = 0.36\pi \frac{v}{\lambda_0} \frac{D}{F} \tag{6.47}$$

It can be seen from Eq. (6.47) that the modulation bandwidth for a diffraction-limited focused Gaussian beam can be very high; for example, for a material of acoustic velocity 5×10^5 cm/sec, the bandwidth of an 0.633 μm light beam focused with an f/10 lens is about 1 GHz. Such a system, however, is practically useless, because the diffraction efficiency would be extremely small.

For the Bragg interaction bandwidth to be large, there must be a large spread in either the acoustic or the optic beam directions $\delta\theta_a$ and $\delta\theta_o$, respectively, or in both. This spread may occur either by focusing, which in the case of the acoustic beam is achieved by curving the plane of the transducer, or it may be due simply to the aperture

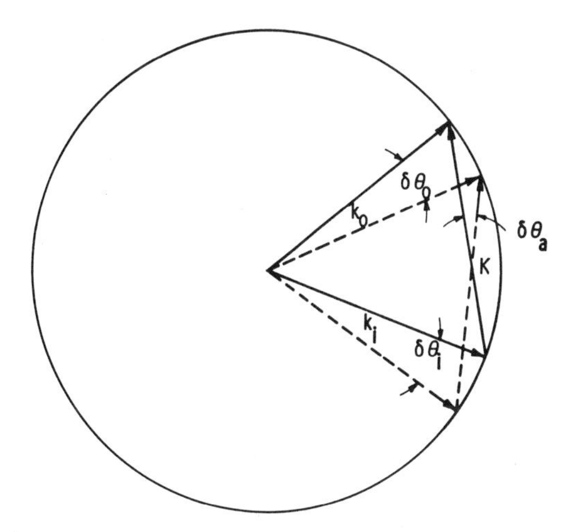

Figure 6.10 Vector diagram for Bragg diffraction in isotropic medium with angular spread of acoustic beam direction.

diffraction for both beams. It follows from fairly simple arguments that the optimum configuration for the most efficient utilization of optic and acoustic energy corresponds to approximately equal angular spreading $\delta\theta_0 \simeq \delta\theta_a$, as illustrated in Fig. 6.10.

To maximize the time-bandwidth product, the angular spread of the acoustic beam should be made large enough to match Bragg diffraction over the frequency range of the transducer-driving circuit bandwidth. As mentioned previously, this will result in some reduction in efficiency. To examine the relationship between bandwidth and the efficiency, we must first state another well-known result of acoustically diffracted light: As shown by Cohen and Gordon [15], the angular distribution of the diffracted light will represent the Fourier transform of the spatial distribution of the acoustic beam. This Fourier transform pair is illustrated in Fig. 6.10 for the usual case of the rectangular acoustic beam profile. It seems intuitively obvious for this simple case, in which the diffraction spread of the incident optic beam is ignored, that there will be components in the diffracted light corresponding to the acoustic field sidelobes. It is shown in Ref. 15 that the Fourier transform relationship holds for an arbitrary acoustic beam profile. For the rectangular profile, the angular dependence of the diffracted light, illustrated in Fig. 6.11, is

$$\frac{I(\theta)}{I_0} \alpha \left[\frac{\sin(1/2)KL(\theta - \theta_B)}{(1/2)KL(\theta - \theta_B)}\right]^2 \tag{6.48}$$

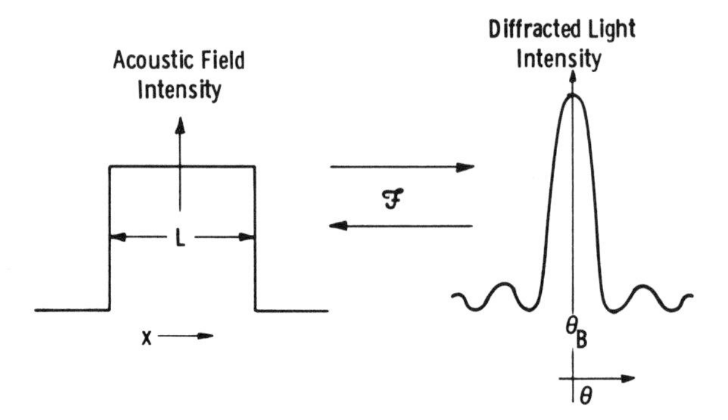

Figure 6.11 Fourier transform relationship between acoustic field intensity and diffracted light intensity.

for which the −3-dB points occur at

$$\frac{1}{2}KL(\Delta\theta)_{1/2} \simeq \pm 0.45\pi \tag{6.49}$$

where $(\Delta\theta)_{1/2}$ is the value of $\theta - \theta_B$ at the half-power points. This yields a value for the angular width of the optic beam just equal to the diffraction spread of the acoustic beam, namely

$$2(\Delta\theta)_{1/2} \simeq \frac{1.8\pi}{KL} \tag{6.50}$$

The frequency bandwidth is obtained by equating this result with the differential of the Bragg condition:

$$\delta\theta = \frac{\lambda_0 \,\Delta f}{nv \cos\theta_B} \tag{6.51}$$

This result is

$$\Delta f = \frac{1.8nv^2 \cos\theta_B}{Lf_0\lambda_0} \tag{6.52}$$

For acousto-optic scanning devices in which the bandwidth as well as the diffraction efficiency is of importance, a more relevant figure of merit may be the product of the bandwidth with the efficiency. By combining Eqs. (6.52) and (6.25), this product is

$$2f_0 \, \Delta f \, \frac{I_1}{I_0} = \frac{1.8\pi^2}{\lambda_0^3 H \cos \theta_B} \left(\frac{n^7 p^2}{\rho v} \right) P \tag{6.53}$$

The quantity in parentheses can be regarded as the figure of merit of the material when the efficiency-bandwidth product is the important criterion and is designated as

$$M_1 = \frac{n^7 p^2}{\rho v} \tag{6.54}$$

Other methods of achieving a large interaction bandwidth include transducer designs that steer the acoustic beam in direction in order to track the Bragg angle as it changes with frequency. A description of beam steering is included in the section on transducers.

Still another figure of merit was introduced by Dixon [16] in connection with wideband acousto-optic devices. Since the power requirements decrease as the transducer height H decreases, it is advantageous to make H as small as possible. If there are no limitations on the minimum size of H, it can be as small as the optic beam waist in the region of the interaction h_{min}. The modulation bandwidth is determined by the travel time of the acoustic wave across this beam waist

$$\tau \approx \frac{1}{\Delta f} = \frac{h_{min}}{v} \tag{6.55}$$

so that

$$h_{min} = \frac{v}{\Delta f} \tag{6.56}$$

Substitution of this value for H in Eq. (6.51) results in the relation

$$2f_0 \frac{I_1}{I_0} = \frac{1.8\pi^2}{\lambda_0^3 \cos \theta_B} \left(\frac{n^7 p^2}{\rho v^2} \right) P \tag{6.57}$$

and the appropriate figure of merit for this situation is the quantity in parentheses:

$$M_3 \equiv \left(\frac{n^7 p^2}{\rho v^2} \right) \tag{6.58}$$

Note that the optic wavelength appears as λ_0^3 in both Eqs. (6.53) and (6.57), so that operation at long wavelengths is relatively more difficult in terms of power requirements for configurations optimizing bandwidth as well as efficiency.

REFERENCES

1. P. Debye and F. W. Sears, *Proc. Nat. Acad. Sci. 18*:409 (1932).
2. R. Lucas and P. Biquard, *J. Phys. Rad. 7th Ser. 3*:464 (1932).
3. C. F. Raman and N. S. N. Nath, *Proc. Indian Acad. Sci. I, 2*:406 (1935).
4. E. I. Gordon, *Proc. IEEE 54*:1391 (Oct. 1966).
5. R. W. Dixon, *IEEE J. Quantum Electron. QE-3*:85 (1967).
6. S. E. Harris, S. T. K. Nieh, and D. K. Winslow, *Appl. Phys. Lett. 15*:325 (1969).
7. R. Mertens, *Meded. K. Vlaam. Acad. Wet. Lett. Schone Kunsten Relg. Kl. Wet. 12*:1 (1950).
8. R. Exterman and G. Wannier, *Helv. Phys. Acta 9*:520 (1936).
9. W. R. Klein and E. A. Hiedemann, *Physica 29*:981 (1963).
10. O. Nomoto, *Jpn. J. Appl. Phys. 10*:611 (1971).
11. W. R. Klein and B. D. Cook, *IEEE Trans. Sonics Ultrason. SU-14*:723 (1967).
12. A. Korpel, *J. Opt. Soc. Am. 69*:678 (1979).
13. A. Korpel and T. Poon, *J. Opt. Soc. Am. 70*:817 (1980).
14. I. C. Chang, *IEEE Trans. Sonics Ultrason. SU-23*:2 (1976).
15. M. Cohen and E. I. Gordon, *Bell Syst. Tech. J. 44*:693 (April 1965).
16. R. W. Dixon, *J. Appl. Phys. 38*:5149 (Dec. 1962).

7

Materials for Acousto-Optic Scanning

7.1 GENERAL CONSIDERATIONS

We have seen in the preceding chapter that two important criteria for choosing materials for acousto-optic scanning systems are the acousto-optic figure of merit and the high-frequency acoustic loss characteristics. Other properties that determine the usefulness of a material are its optic transmission range, optic quality, availability in suitable sizes, mechanical and handling characteristics as they may pertain to polishing and fabrication procedures, and chemical stability under normal conditions. As with most components, cost will be an important factor, even when all the other factors may be positive, if competing techniques are available.

One of the limitations on the use of acousto-optic scanners before the late 1960s was the availability of materials with a reasonably high figure of merit. As we have seen, fused quartz, which is used as the standard for comparison, has a figure of merit so low that only a few percent diffraction efficiency can be obtained for scanners of typical dimensions and with RF powers that can be applied without causing damage to transducer structures. Water is a fairly efficient material, with a figure of merit about 100 times larger than fused quartz, and has actually found use in some scanning systems. Like most liquids, it cannot be used at frequencies higher than about 50 MHz, so that large numbers of resolution elements cannot be achieved. Since the late 1960s, many new materials have been synthesized and existing ones discovered to have excellent properties. Materials can now be found for most scanning applications from the UV through the intermediate IR where high bandwidth is required.

The selection of a material for any particular device will be dictated by the type of operation under consideration. In general, it is desirable to select a material with low drive power requirements, suggesting those with large refractive index, and low density and acoustic velocity. If, however, high speed modulation is of paramount importance, then a low acoustic velocity may lead to slower than required speeds. In the following section, we shall review the factors and tradeoffs involved in the selection of materials for various acousto-optic applications. Whatever the particular material requirements may be, there are also a number of practical considerations which dictate several generally important material properties whatever the application: (1) The optic quality must be high so that not only absorption but scattering and large-scale inhomogeneities are small; (2) good chemical stability is required so that protective enclosures are not needed to maintain integrity; (3) good mechanical properties are required so that the device can be cut and polished without extraordinary procedures and can be adjusted and used with normal handling techniques; (4) the availability of crystal growth methods for obtaining suitably large, high-quality boules with reasonable cost is necessary; and (5) a low-temperature coefficient of velocity is required to avoid drift of scan properties.

7.2 THEORETICAL GUIDELINES

There is no simple microscopic theory of the photoelastic effect in crystals. Therefore it is not possible to predict the magnitude of the photoelastic constants from first principles. However, Pinnow [1] has suggested the use of certain empirical relationships between the various physical properties in order to systemitize and group acousto-optic materials. It is well known that such relations exist, for example, for the refractive index and the acoustic velocity for such groups as the alkali halides, the mineral oxides, and the III-IV compounds.

A great amount of data has been collected on the refractive indices of crystals, and generally good agreement is found with the Gladstone-Dale [2] equation

$$\frac{n-1}{\rho} = \sum_i q_i R_i \tag{7.1}$$

in which R_i is the specific refraction of the ith component and q_i is percentage by weight. Reliable values of R_i have been determined from mineralogical data over many years.

From the expression for the acousto-optic figure of merit, it is apparent that a high value of refractive index is desirable for achieving high diffraction efficiency. It is not, however, possible simply to select for consideration those materials with high refractive index, as even a casual survey shows that such materials tend to be opaque at shorter wavelengths. This trend was examined in great detail by Wemple and DiDomenico [3], who found that the refractive index is simply related to the energy band gap. The semiempirical relation for oxide materials is

$$n^2 = 1 + \frac{15}{E_g} \tag{7.2}$$

where E_g is the energy gap (expressed in electronvolts). For other classes of materials the energy gap constant will be different, but the same form holds. It can be seen from Eq. (7.2) that the largest refractive index for an oxide material transparent over the entire visible range (cutoff wavelength at 0.4 μm) is 2.44. Higher refractive indices can be chosen only by sacrificing transparency at short wavelengths.

Pinnow [1] has found that a good approximation to the acoustic velocity for a wide range of materials is obtained with the relation

$$\log\left(\frac{v}{\rho}\right) = -b\overline{M} + d \tag{7.3}$$

where $\overline{M}$ is the mean atomic weight, defined as the total molecular weight divided by the number of atoms per molecule, and b and d are constants. Large values of d are generally associated with harder materials, while b does not vary greatly for oxides. Thus, in general, low acoustic velocities tend to be found in materials of high density, as is intuitively expected. Another useful velocity relationship has been pointed out by Uchida and Niizeki [4]; this is the Lindemann formula relating the melting temperature T_m and the mean acoustic velocity v_m

$$v_m^2 = \frac{cT_m}{\overline{M}} \tag{7.4}$$

in which c is a constant, dependent on the material class. This relation suggests that high-efficiency materials would likely be found among those with large mean atomic weight and low melting temperature, that is, dense, soft materials.

For an acoustic-optic material to be useful for wideband applications, the ultrasonic attenuation must be small at high frequencies. An attenuation that is often taken as an upper limit is 1 dB/μsec (so that the useful aperture will depend on the velocity). Many materials that might be highly efficient and otherwise suitable are excessively lossy at high frequency. A microscopic treatment of ultrasonic attenuation was done by Woodruff and Ehrenreich [5]. Their formula for the ultrasonic attenuation is

$$\alpha = \frac{\gamma^2 \Omega^2 \kappa T}{\rho v^5} \tag{7.5}$$

where Ω is the radian frequency, γ is the Grüneisen constant, κ is the thermal conductivity, and T is the absolute temperature. This formula would suggest that the requirement of low acoustic velocity and low attenuation conflict with each other, since $\alpha \sim v^{-5}$; it is very exceptional that materials with low acoustic velocity do not also have a high absorption, at least for the low-velocity modes.

Table 7.1 Maximum Photoelastic Coefficients

Material	$\lvert p_{max} \rvert_{measured}$
$LiNbO_3$	0.20
TiO_2	0.17
Al_2O_3	0.25
$PbMoO_4$	0.28
TeO_2	0.23
$Sr_{0.5}Ba_{0.5}Nb_2O_6$	0.23
SiO_2	0.27
YIG	0.07
$Ba(NO_3)_2$	0.35
α-HIO_3	0.50
$Pb(NO_3)_2$	0.60
ADP	0.30
CdS	0.14
GaAs	0.16
As_2S_3	0.30

Source: From Ref. 9.

The determination of the photoelastic constants of materials is essentially an empirical study, although a microscopic theory of Mueller [6], developed for cubic and amorphous structures, is still referenced. For both ionic and covalent bonded materials the photoelastic effect derives from two mechanisms, the change of refractive index with density and the change in index with polarizability under the strain. Both of these effects may have the same or opposite sign under a given strain, and one or the other may be the larger. It is for this reason that the magnitude or even the sign of the photoelastic constant cannot be predicted, since the effects may completely cancel each other. It is possible, however, to estimate the maximum constants for groups of materials. This has been done for three important groups with the result

$$|p_{max}| = \begin{cases} 0.21 & \text{(water-insoluble oxides)} \\ 0.35 & \text{(water-soluble oxides)} \\ 0.20 & \text{(alkali halides)} \end{cases}$$

In general, the photoelastic tensor components corresponding to shear strain will be less than those corresponding to compressional strain because there is no change, to first order, of density with shear; only the polarizability effect will be present. It is always possible that exceptionally large values of shear-related photoelastic coefficients may be found, but in no case could they be expected to be larger than the estimated value of $|p_{max}|$. The maximum values of photoelastic constant are shown in Table 7.1 for a number of important oxides and other materials.

7.3 SELECTED ACOUSTO-OPTIC MATERIALS

Among older materials, those that have been shown useful for acousto-optic applications are fused quartz because of its excellent optic quality and low cost for large sizes and sapphire and lithium niobate because of their exceptionally low acoustic losses at microwave frequencies. For infrared applications germanium [7] has proven very useful, as has arsenic trisulfide glass, where bandwidth requirements are not high. Among the newer crystal materials, very good acousto-optic performance has been obtained in the visible range with α-HIO_3 [8] and $PbMoO_4$ [9,10]. One of the most interesting new materials to be developed within the past several years is TeO_2 [11], which along with $PbMoO_4$ has found wide use in commercially available acousto-optic scanners. More design details for devices employing this material will be given later. Among the new materials that have been developed for infrared applications, very high performance has been reached with several chalcogenide crystals [12]. Particularly important members of this group of materials are Tl_3AsS_4 [3] and Tl_3PSe_4 [14]. The compound Tl_3AsSe_3 [15] is particularly interesting beyond its possible use as an infrared acousto-optic modulator material. Since Tl_3AsSe_3 belongs to the crystal class 3 m, its symmetry permits it to possess a nonzero p_{41} photoelastic coefficient, and it is suitable for use as a collinear tunable acousto-optic filter, a device first realized by Harris [16] using lithium niobate.

The following tables summarize the properties of some of the materials that have been studied for acousto-optic applications. The acoustic attenuation constant in these tables is defined as

$$\Gamma = \frac{\alpha}{f^2} \qquad (7.6)$$

which supposes that the attenuation increases quadratically with frequency. This will be the case for good-quality single crystals but not for polycrystalline, highly impure, or amorphous materials. For the latter, the constant given in the tables is a rough estimate, based

Table 7.2 Acousto-optic Properties of Amorphous Materials

Material	Trans. range (μm)	Acoustic mode	v (cm/sec × 10^5)	Γ (dB/cm GHz^2)	Opt. pol. dir.[b]	n (0.633 μm)	M_1 (cm^2 sec/g × 10^{-7})	M_2 (sec^3/g × 10^{-18})	M_3 (cm sec^2/g × 10^{-12})
Water	0.2-0.9	L[a]	1.49	2400	‖ or ⊥	1.33	37.2	126	25
Fused quartz	0.2-4.5	L	5.96	12	⊥	1.46	8.05	1.56	1.35
SF-4	0.38-1.8	L	3.63	220	⊥	1.62	1.83	4.51	3.97
SF-59	0.46-2.5	L	3.20	1200	‖ or ⊥	1.95	39	19	12
SF-58		L	3.26	1200	‖ or ⊥	1.91	18.2	9	5.6
SF-57		L	3.41	500	‖	1.84	19.3	9	5.65
SF-6		L	3.51	500	‖ or ⊥	1.80	15.5	7	4.42
As_2S_3	0.6-11	L	2.6	170	‖	2.61	762	433	293
As_2S_5	0.5-10	L	2.22			2.2[c]	278[c]	256[c]	125

[a]Longitudinal.
[b]With respect to acoustic wave propagation direction.
[c]Estimated.

Table 7.3 Acousto-optic Properties of Crystals for the Visible

Material	Trans. range (μm)	Acoustic mode & prop. dir.	v (cm/sec $\times 10^5$)	Γ (dB/cm GHz^2)	Opt. pol. dir.	n (0.633 μm)	M_1 (cm^2 sec/g $\times 10^{-7}$)	M_2 (sec^3/g $\times 10^{-18}$)	M_3 (cm sec^2/g $\times 10^{-12}$)
$LiNbO_3$	0.4-4.5	L[100]	6.57	0.15		2.20	66.5	7.0	10.1
		S[001]	3.59	2.6	$\perp$	2.29	9.2	2.92	2.4
Al_2O_3	0.15-6.5	L[100]	11.0	0.2	$\parallel$	1.77	7.7	0.36	0.7
YAG	0.3-5.5	L[110]	8.60	0.25	$\perp$	1.83	0.98	0.073	0.114
		S[100]	5.03	1.1	$\parallel$ or $\perp$	1.83	1.1	0.25	0.23
TiO_2	0.45-6	L[001]	10.3	0.55	$\perp$	2.58	44	1.52	4
SiO_2	0.12-4.5	L[001]	6.32	2.1	$\perp$	1.54	9.11	1.48	1.44
		L[100]	5.72	3.0	[001]	1.55	12.1	2.38	2.11
α-HIO_3	0.3-1.8	L[001]	2.44	10	[100]	1.99	103	86	42
$PbMoO_4$	0.4-4	L[001]	3.63	15	$\parallel$	2.62	108	36.3	29.8
TeO_2	0.35-4	L[001]	4.20	15	$\perp$	2.26	138	34.5	32.8
		S[110]	0.616	90	Circ. [001]	2.26	68.0	793	110
Pb_2MoO_5	0.4-4	L, a axis	2.96	25	b axis	2.183	242	127	82

Table 7.4 Acousto-optic Properties of Infrared Crystals

Material	Trans. range (μm)	Acoustic mode & prop. dir.	v (cm/sec × 10^5)	Γ (dB/cm GHz^2)	Opt. pol. dir.	λ (μm)	n	M_1 (cm^2 sec/g × 10^{-7})	M_2 (sec^3/g × 10^{-18})	M_3 (cm sec^2/g × 10^{-12})
Ge	2-20	L[111]	5.50	30	∥	10.6	4.00	10,200	840	1850
		S[100]	3.51	9	∥ or ⊥	10.6	4.00	1,430	290	400
Tl_3AsS_4	0.6-12	L[001]	2.5	29	∥	1.15	2.63	620	510	290
GaAs	1-11	L[110]	5.15	30	∥	1.15	3.37	925	104	179
		S[100]	3.32		∥ or ⊥	1.15	3.37	155	46	49
Ag_3AsS_3	0.6-13.5	L[001]	2.65	800	∥	0.633	2.98	816	390	308
Tl_3AsSe_3	1.25-18	L[100]	2.15	314	⊥	3.39	3.15	654	445	393
Tl_3PSe_4	0.85-9	L[100]	2.0	150	∥	1.15	2.9	2,866	2069	1238
$TlGaSe_2$	0.6-20	L[001]	2.67	240	∥	0.633	2.9	430	393	161
CdS	0.5-11	L[100]	4.17	90	∥	0.633	2.44	52	12	12
ZnTe	0.55-20	L[110]	3.37	130	∥	1.15	2.77	75	18	19
GaP	0.6-10	L[110]	6.32	6.0	∥	0.633	3.31	75	30	71
ZnS	0.4-12	L[001]	5.82	27	∥	0.633	2.35	27	3.4	4.7
		S[001]	2.63	130		0.633	2.35	14	8.4	5.2
Te	5-20	L[100]	2.2	60	∥	10.6	4.8	10,200	4400	4640

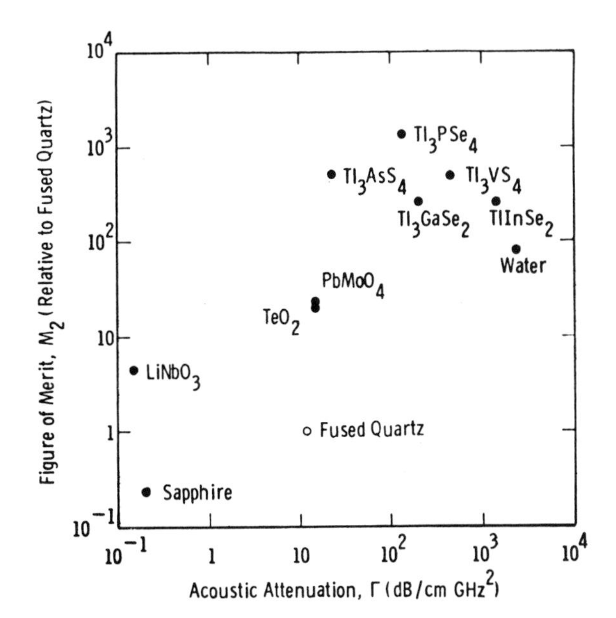

Figure 7.1 Figure of merit vs. acoustic attenuation.

on measurements at the higher frequencies. The light polarization direction is designated as parallel or perpendicular according to whether the light polarization is parallel or perpendicular to the acoustic beam direction. Table 7.2 lists some of the more important amorphous materials, which may be useful if large sizes are desired, but none of which can be used at frequencies much above 30 MHz. Table 7.3 lists the most important class of materials, crystals that are transparent throughout the visible and with very low acoustic losses. Table 7.4 lists a number of high-efficiency crystal materials that are transparent in the infrared and that have reasonably low acoustic losses.

An overall summary of the best of the selected acousto-optic materials presented in these tables is shown in Fig. 7.1. Using figure of merit and acoustic attenuation as criteria of quality, it is clear that a trade-off between these two exists and that the selection of the optimum material will be determined by the system requirements.

7.4 ANISOTROPIC DIFFRACTION IN TELLURIUM DIOXIDE

One of the most remarkable materials to have appeared recently for acousto-optic applications is paratellurite, TeO_2 [17]. It has a unique combination of properties which leads to an extraordinarily high figure of merit for a shear wave interaction in a convenient RF range. It will be recalled that the anisotropic Bragg relations, Eq. (6.30) and (6.31), led to a particular frequency, given by Eq. (6.35), for which the angle of incidence is a minimum and therefore satisfies the Bragg condition over a wide frequency range. However, typical values of birefringence place this frequency around 1 GHz or higher. Of particular interest in TeO_2 is its optic activity for light propagating along the c axis or [001] direction; the indices of refraction for left- and right-hand circularly polarized light are different so that plane polarized light undergoes a rotation of its plane of polarization by an amount

$$R = \frac{2n_o}{\lambda}\delta \tag{7.7}$$

where δ is the index splitting between left- and right-hand polarized light:

$$\delta = \frac{n_\ell - n_r}{2n_o} \tag{7.8}$$

Just as acoustic shear waves can phase match two linearly polarized light waves, they can also phase match two oppositely circularly polarized light waves. Thus, shear waves propagating in the [110] direction, with shear polarization in the $[\bar{1}10]$ direction, will diffract left- and right-hand polarized light propagating along the [001] direction into each other. The anisotropic Bragg relations apply to crystals with optic activity, where the birefringence is interpreted as

$$\Delta n = n_\ell - n_r = 2n_o\delta \tag{7.9}$$

and the value of δ obtained from specific rotation is wavelength dependent. For the light and sound wave propagation directions described above, the acoustic velocity is 0.62×10^5 cm/sec, and the figure of merit M_2 is 515 relative to fused quartz. The frequency for which the Bragg angle of incidence is a minimum, as evaluated from Eq. (6.35) for $\lambda = 0.633$ μm, is f = 43 MHz, a very convenient frequency. For other important wavelengths, the minima occur at 60 MHz for $\lambda = 0.5145$ μm and at 15 MHz for $\lambda = 1.15$ μm.

The application of the anisotropic Bragg equations to optically active crystals was discussed in detail by Warner et al. [18]. They showed that near the optic axis the indices of refraction are approximated by the relation (for right-handed crystals, $n_r < n_\ell$)

$$\frac{n_r^2(\theta)\cos^2\theta}{n_o^2(1-\delta)^2} + \frac{n_r^2(\theta)\sin^2\theta}{n_e^2} = 1 \tag{7.10}$$

and

$$\frac{n_\ell^2(\theta)\cos^2\theta}{n_o^2(1+\delta)^2} + \frac{n_\ell^2(\theta)\sin^2\theta}{n_o^2} = 1 \tag{7.11}$$

For incident angles near zero with respect to the optic axis and for small values of δ ,

$$n_r^2 = n_o^2\left(1 - 2\delta + \frac{n_e^2 - n_o^2}{n_e^2}\sin^2\theta\right) \tag{7.12}$$

and

$$n_\ell^2 = n_o^2(1 + 2\delta\cos^2\theta) \tag{7.13}$$

For light incident exactly along the optic axis the two refractive indices are simply

$$n_r = n_o(1 - \delta) \tag{7.14}$$

and

$$n_\ell = n_o(1 + \delta) \tag{7.15}$$

The anisotropic Bragg equations for optically active crystals are obtained by substitution of Eqs. (7.12) and (7.13) into Eqs. (6.30) and (6.31) for n_i and n_d. By ignoring the higher-order terms, this results in

$$\sin\theta_r \simeq \frac{\lambda f}{2n_o v}\left[1 + \frac{4n_o^2v^2}{\lambda^2f^2}\delta + \frac{\sin^2\theta\, n_r^2 n_o^2}{\lambda^2f^2}\left(\frac{n_e^2 - n_o^2}{n_e^2}\right)\right] \tag{7.16}$$

and

$$\sin\theta_e = \frac{\lambda f}{2n_o v}\left[1 - \frac{4n_o^2v^2}{\lambda^2f^2}\delta - \frac{\sin^2\theta\, n_e^2 n_o^2}{\lambda^2f^2}\left(\frac{n_e^2 - n_o^2}{n_o^2}\right)\right] \tag{7.17}$$

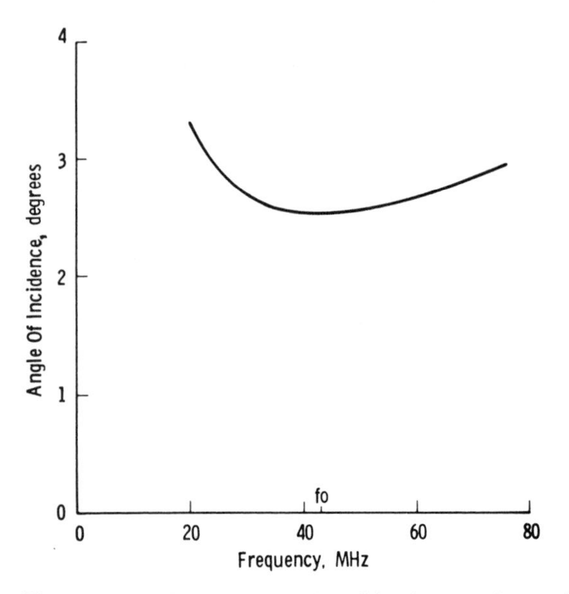

Figure 7.2 External angle of incidence for anisotropic bragg diffraction in TeO_2 at $\lambda = 0.6328\ \mu m$.

The angle of incidence (as measured external to the crystal) is shown in Fig. 7.2 for TeO_2 at $\lambda = 0.6328\ \mu m$. It is obvious that for frequencies around the minimum it will be possible to achieve a much larger bandwidth for a given interaction length than is possible with normal Bragg diffraction; a one-octave bandwidth corresponds to a variation in angle of incidence for perfect phase matching of only 0.16°. A useful advantage of such operation is that large bandwidths are compatible with large interaction lengths, which assumes the avoidance of higher diffraction orders from Raman-Nath effects. For normal Bragg diffraction, on the other hand, large bandwidths can only be reached with interaction lengths that are so small that significant higher-order diffraction occurs. This decreases the efficiency with which light can be directed to the desired first order and also limits the bandwidth to less than one octave in order to avoid overlapping low-frequency second-order with higher-frequency first-order diffracted light. However, tellurium dioxide operating in the anisotropic mode can always be made to diffract in the Bragg mode with no bandwidth limitation on length, so that this feature combined with the extraordinary high figure of merit leads to deflector operation with very low drive powers.

An important degeneracy occurs for anisotropic Bragg diffraction which causes a pronounced dip in the diffracted light intensity at the midband frequency, where θ_i has its minimum. This degeneracy was explained by Warner et al. [18] and is easily understood by referring

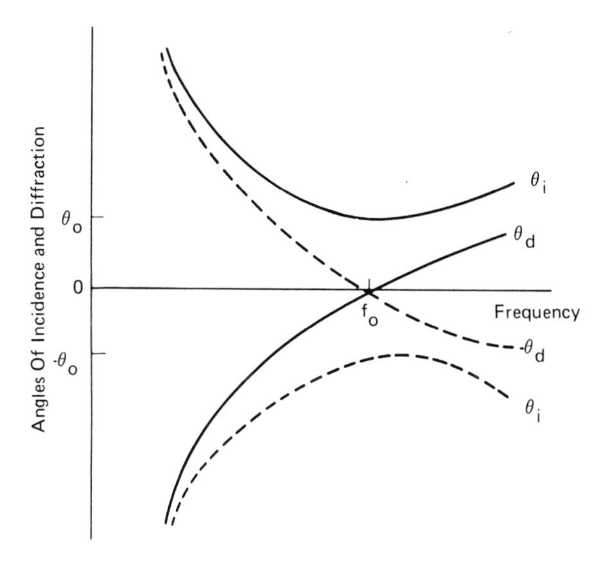

Figure 7.3 Angles of incidence and diffraction for anisotropic diffraction. Solid curves are for incident light having a component in the same direction as the acoustic wave, and dotted curves are for incident light having a component in the opposite direction.

to the diagram in Fig. 7.3. Two sets of curves are shown in the figure; the solid pair represents θ_i and θ_d when the incident light momentum vector has a positive component along the acoustic momentum vector, and the dotted pair represents these angles when the incident light momentum vector has a negative component along the acoustic vector. In the former case, the frequency of the diffracted light is upshifted, and in the latter is is downshifted. The vector diagram for this process is shown in Fig. 7.4. Light is incident to the acoustic wave of frequency f_0 at an angle θ_0 and is diffracted as a frequency-

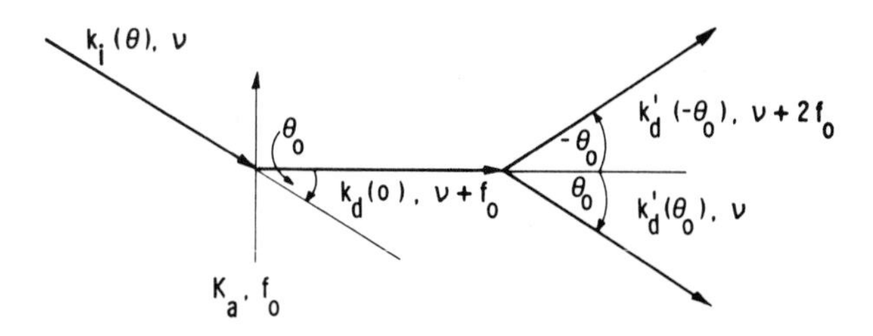

Figure 7.4 Vector diagram for midband degeneracy of Bragg diffraction in birefringent medium.

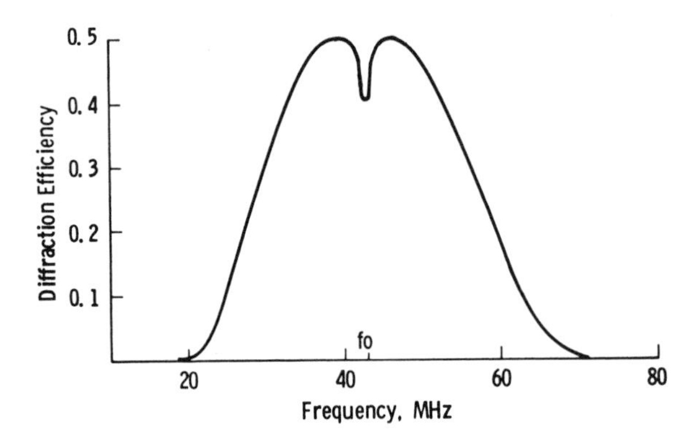

Figure 7.5 Effect of midband degeneracy on diffraction efficiency for a maximum efficiency of 50%.

upshifted beam $\nu + f_0$ normally to the acoustic wave. This light, in turn, may be rediffracted; referring to Fig. 7.3, it can be seen that for a frequency of f_0, light that is incident at $\theta = 0°$ can be rediffracted to either θ_0 or $-\theta_0$. In the former case, the light will be downshifted to the original incident light frequency ν, and in the latter case it is upshifted to $\nu + 2f_0$. Note that this degeneracy can only occur at the frequency f_0 where light incident normally to the acoustic wave is phase-matched for diffraction into both θ_0 and $-\theta_0$. How the light is distributed in intensity between the three modes will depend on the interaction length and the acoustic power level. The exact solution to this is found by setting up the coupled mode propagation equations under phase-matched conditions. The result of this is that maximum efficiency for deflection into the desired mode at f_0 is 50%. At low acoustic power the deflection of light into the undesired mode is negligible; at high powers the unwanted deflection increases so that, for example, if the efficiency is 50% for frequencies away from f_0, it will be 40% at f_0. The theoretical response of such a deflector is shown in Fig. 7.5 and is in excellent agreement with experimental results.

REFERENCES

1. D. A. Pinnow, *IEEE J. Quantum Electron. QE-6*:223 (April 1970).
2. J. H. Gladstone and T. P. Dale, *Philos. Trans. R. Soc. London 153*:37 (1964).
3. S. H. Wemple and M. DiDomenico, *J. Appl. Phys. 40*:735 (1969).
4. N. Uchida and N. Niizeki, *Proc. IEEE 61*:1073 (1973).

5. T. O. Woodruff and H. Ehrenreich, *Phys. Rev.* *123*:1553 (1961).
6. H. Mueller, *Phys. Rev.* *47*:947 (1935).
7. R. L. Abrams and D. A. Pinnow, *J. Appl. Phys.* *41*:2765 (1970).
8. D. A. Pinnow and R. W. Dixon, *Appl. Phys. Lett.* *13*:156 (1968).
9. D. A. Pinnow, L. G. Van Uitert, A. W. Warner, and W. A. Bonner, *Appl. Phys. Lett.* *15*:83 (1969).
10. G. A. Coquin, D. A. Pinnow, and A. W. Warner, *J. Appl. Phys.* *42*:2162 (1971).
11. Y. Ohmachi and N. Uchida, *J. Appl. Phys.* *40*:4692 (1969).
12. M. Gottlieb, T. J. Isaacs, J. D. Feichtner, and G. W. Roland, *J. Appl. Phys.* *40*:4692 (1969).
13. G. W. Roland, M. Gottlieb, and J. D. Feichtner, *Appl. Phys. Lett.* *21*:52 (1972).
14. T. J. Isaacs, M. Gottlieb, and J. D. Feichtner, *Appl. Phys. Lett.* *24*:107 (1974).
15. J. D. Feichtner and G. W. Roland, *Appl. Opt.* *11*:993 (1972).
16. S. E. Harris and R. W. Wallace, *J. Opt. Soc. Am.* *59*:744 (1969).
17. N. Uchida and Y. Ohmachi, *J. Appl. Phys.* *40*:4692 (1969).
18. A. W. Warner, D. L. White, and W. A. Bonner, *J. Appl. Phys.* *43*:4489 (Nov. 1972).

8
Acoustic Techniques

8.1 TRANSDUCER CHARACTERISTICS

The second key component of the acousto-optic scanner after the optical medium is the transducer structure, which includes the piezoelectric layer, bonding films, backing layers, and matching network. Recent advances in this area have made available a number of new piezoelectric materials of very high electromechanical conversion efficiency and bonding techniques that permit this high conversion efficiency to be maintained over a large bandwidth. Furthermore, the design of high-performance transducer structures utilizing this new technology has been facilitated by new analytic tools [1,2] which lend themselves to computer programs for optimizing this performance.

The most elementary configuration of a thickness-driven transducer structure is shown in Fig. 8.1. It consists of the piezoelectric layer, thin film or plate, excited by metallic electrodes on both faces and a bonding layer to acoustically couple the piezoelectric to the delay medium. The backing is applied to mechanically load the transducer for bandwidth adjustment but may simply be left as air. The thickness of the transducer is one-half an acoustic wavelength at the resonant frequency, and the thickness of the bonding layer is ideally kept negligibly small compared with an acoustic wavelength. The most efficient operation of the transducer is obtained when the mechanical impedance of all the layers are equal. The mechanical impedance is

$$Z = \rho v \tag{8.1}$$

and in general there is not sufficient choice of available materials to satisfy this condition. When the impedances are unequal, reflection occurs at the interface, reducing the efficiency of energy transfer. The reflection and transmission coefficients at the boundary between two media of impedances Z_1 and Z_2 are

$$R = \frac{(Z_1 - Z_2)^2}{(Z_1 + Z_2)^2} \tag{8.2}$$

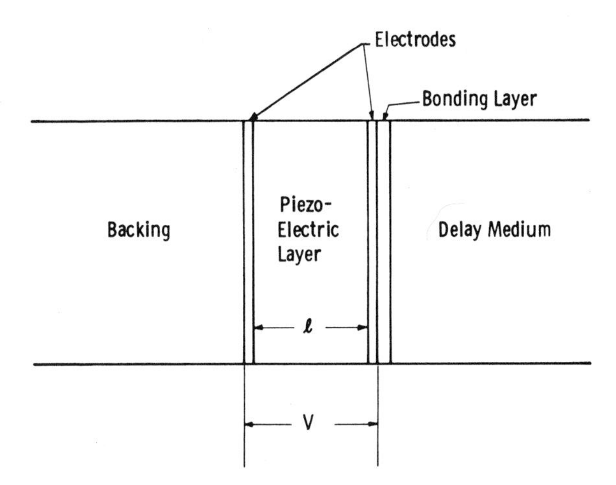

Figure 8.1 Transducer structure.

$$T = \frac{4Z_1Z_2}{(Z_1 + Z_2)^2} \tag{8.3}$$

The electromechanical analysis is generally carried out in terms of an equivalent circuit model, first proposed by Mason [3]. Several variations of the equivalent circuit have since been developed, but the one due to Mason is shown in Fig. 8.2. The fundamental constants of the transducer are permittivity ε, acoustic velocity v, and electromechanical coupling factor κ. The other parameters are transducer thickness ℓ and area S. With these parameters, the circuit components shown in Fig. 8.2 are

$$C_0 = \varepsilon \frac{S}{\ell} \tag{8.4}$$

$$\phi = \kappa \left(\frac{1}{\pi} \omega_0 c_0 Z_0 \right)^{1/2} \tag{8.5}$$

$$Z_A = jZ_0 \tan \frac{\gamma}{2} \tag{8.6}$$

$$Z_B = -j \frac{Z_0}{\sin \gamma} \tag{8.7}$$

where

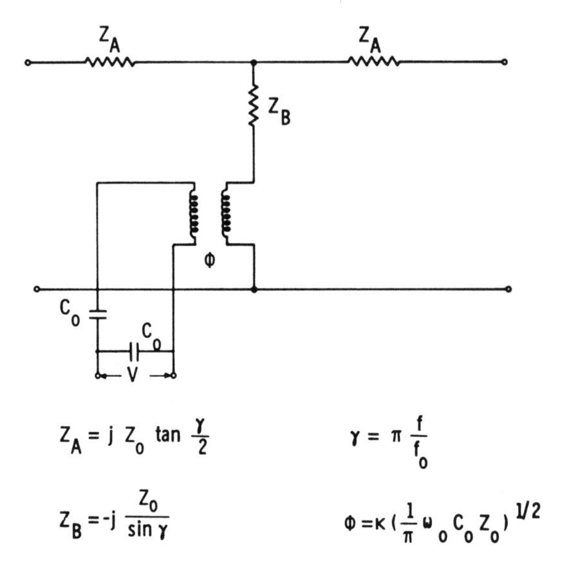

Figure 8.2 Equivalent circuit of Mason.

$$\omega_0 = \frac{\pi v}{\ell} \tag{8.8}$$

$$\gamma = \pi \frac{\omega}{\omega_0} \tag{8.9}$$

$$Z_0 = S\rho v \tag{8.10}$$

This equivalent circuit was used by Sittig [1] and Meitzler and Sittig [2] to analyze the propagation characteristics of acoustic energy between a piezoelectric and a delay medium. This was done in terms of a two-port electromechanical network, described by the chain matrix

$$\begin{bmatrix} A & B \\ C & D \end{bmatrix} = \prod_m \begin{bmatrix} A_m & B_m \\ C_m & D_m \end{bmatrix} \tag{8.11}$$

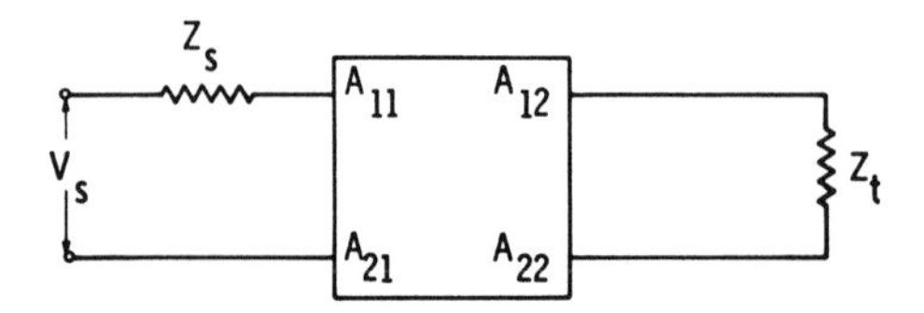

Figure 8.3 Terminated two-port transducer.

If the equivalent circuit of Fig. 8.2 is terminated at the input with a source of voltage V_s and impedance Z_s and at the output with a transmission medium of mechanical impedance Z_t and output voltage and load impedance V_ℓ and Z_ℓ, as shown in Fig. 8.3, then the insertion loss is

$$L = 20 \log \frac{V_s}{V_\ell} + 20 \log \left| \frac{Z_s + Z_\ell}{Z_\ell} \right| \text{ dB} \tag{8.12}$$

The impedance Z_s and Z_ℓ are assumed to be purely resistive and

$$\frac{V_\ell}{V_s} = \frac{2Z_\ell Z_t}{\{[AZ_t + B + Z_s(CZ_t + D)][AZ_t + B + Z_\ell(CZ_t + D)]\}} \tag{8.13}$$

The two-port transfer matrix was obtained by Sittig and Cook [4] with the result

$$A = \frac{1}{\phi H}\begin{bmatrix} A' & B' \\ C' & D' \end{bmatrix}\begin{bmatrix} \cos\gamma + jz_b \sin\gamma & Z_0(z_b \cos\gamma + z \sin\gamma) \\ (j \sin\gamma)/Z_0 & 2(\cos\gamma - 1) + jz_b \sin\gamma \end{bmatrix} \tag{8.14}$$

where

$$z_b = \frac{Z_b}{Z_0}, \quad H = \cos\gamma - 1 + jZ_b \sin\gamma \tag{8.15}$$

and

$$A' = 1, \quad B' = j\frac{\phi^2}{\omega C_0}, \quad C' = j\omega C_0, \quad D' = 0 \tag{8.16}$$

The impedance Z_b represents the mechanical impedance of layers placed on the back surface of the transducer for loading, $Z_b = S\rho_b v_b$. In case the transducer is simply air backed, $Z_b \simeq 0$. Electrical matching may be done at the input network by adding inductors either in parallel or in series in order to be electrically resonant with the transducer capacity C_0 at midband $\omega = \omega_0$. If no inductances are added, the minimum loss condition is achieved for

$$R_s = \frac{1}{\omega_0 C_0} \tag{8.17}$$

where R_s is the source resistance. The inductance, if added, is chosen so that

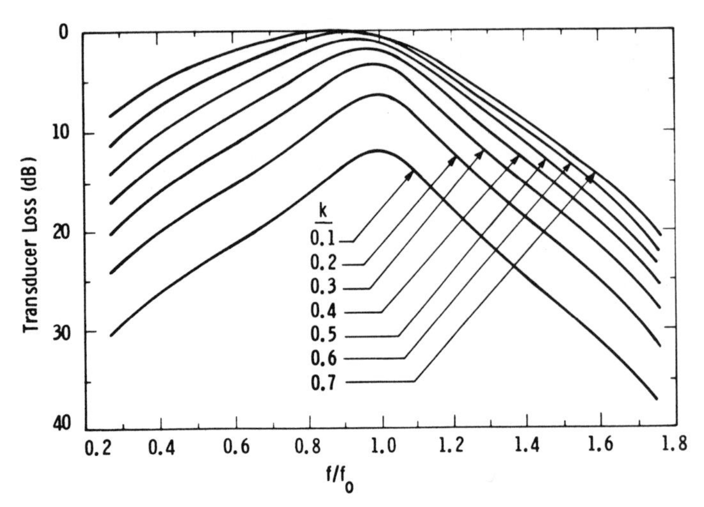

Figure 8.4 Transducer loss for various values of κ; $z_{0t} = 0.4$ and $R_s = (\omega_0 C_0)^{-1}$.

$$L = \frac{1}{\omega_0^2 C_0} \tag{8.18}$$

A result of the matrix analysis shows that when piezoelectric materials with large values of the coupling constant κ are used, it is possible to achieve large fractional bandwidths without the necessity for electrical matching networks. As an example of the results obtained with this formalism, several plots of the frequency dependence of transducer loss for different values of the coupling constant are shown in Fig. 8.4.

8.2 TRANSDUCER BONDING

For transducers in the frequency range for which crystal plates are bonded to the delay medium, the bonding procedure is probably the most critical and most difficult step in fabricating the structure. The bonding layer can drastically modify the transmission of acoustic energy between the piezoelectric and the delay medium, because the bond layer must provide molecular contact between the two surfaces, which will otherwise result in incomplete transfer, and because the mechanical impedance of the bond layer may produce a serious mismatch, with low transmission. In addition to these considerations, if the bond material is acoustically lossy, further decrease in transmission will result. Because of the special properties required, there is only a very limited number of known bonding materials available. For temporary attachments, a commonly used agent is *salol*, phenyl salicylate. It is easily applied as a liquid, which is crystallized by addition of a small seed. It is reliquefied by gentle heating and therefore is useful for various test measurements but does not yield wide bandwidth or efficient coupling. A more satisfactory bond is made with epoxy resin, mixed to a very low viscosity which may be compressed to a layer less than 1 μm thick before setting. Such thin layers require a high degree of cleanliness to avoid inclusion of any dust particles. Because of the low impedance of epoxy compared with such transducer materials as lithium niobate, thicker bonding layers would cause serious impedance mismatch problems in the frequency range around 100 MHz, where this technique has been successfully used. For frequencies higher than about 100 MHz, other techniques, capable of yielding still thinner bond layers, which must be kept to a small fraction of an acoustic wavelength, must be used. Vacuum-deposited metallic layers are well suited for this purpose, since their thickness can be very accurately controlled down to the smallest dimensions, and impedances much closer to those of commonly used piezoelectric materials are available. The best results have been obtained with indium bonds [5], which are deposited to a thickness of several thousand angstroms on both surfaces and, without removal from the vacuum systems, are mated under a pressure of about 1000 psi. This technique yields a cold-welded bond which has excellent mechanical properties with large acoustic bandwidth if properly designed and low insertion loss at frequencies of hundreds of megahertz. The greatest fabrication difficulty is due to the necessity of maintaining the deposited films under vacuum to prevent oxidation. This requires a vacuum system with rather elaborate fixtures to bring the two surfaces together after film deposition and to apply the hydraulic pressure. In a modification of the indium bond [6] which allows the freshly deposited indium surfaces to be removed from the vacuum system for handling, the work is then placed in an oven under a pressure of several hundred psi and raised in temperature to slightly below the melting point of indium (156°C) and then slowly cooled. This procedure forms a molecular bond in spite of the oxidation that may occur and gives results similar to the vacuum bond. The principal drawback is that upon cooling differential thermal expansion coefficients between the delay line material and the transducer material may set up unacceptable strains in the optic path. For some systems, this may not be a problem; for example, quartz or even lithium niobate transducers on fused or crystal quartz delay lines can be routinely made by this method. On the other hand, such crystals as tellurium dioxide require a great deal of care in handling, since they are extremely sensitive to thermal shock and strain. Differential contraction between the crystal and transducer for a bond made in this fashion may easily be severe enough to fracture the crystal. Therefore, its applicability will depend on the materials and sizes involved and on the degree of freedom from residual strain required.

For frequencies approaching 1 GHz, the attenuation of indium layers may become excessive, and better results can be achieved with metals with lower acoustic loss constants. Among such metals are gold, silver, and aluminum. These can be cold-welded in vacuum, as indium, but generally require higher pressures. Still another method that has been used with these, as well as indium, is ultrasonic welding [7]. The chief advantage to be gained is that the procedure is carried out in normal atmosphere, since the ultrasonic energy breaks up the oxidation layer that forms on the surface. Some heating occurs as a result, but the temperature remains well below that required in the indium thermocompression method, with much lower residual strain. The technique requires the simultaneous application of pressures up to 3000 psi; this may be excessive for easily fractured or deformed materials or where odd-shaped samples are involved. A summary of the important properties of a few bonding materials, also used for electrodes and intermediate impedance matching layers, is given in Table 8.1.

At lower frequencies, the effects of thin electrode and bonding layers on the performance of the transducer may be entirely negligible, but in the range near 100 MHz, they become increasingly large, and even for layers less than 1 μm in thickness the effect may not be negligible if the impedance mismatch to the rest of the structure is large. The effects of the electrode layer can be determined by setting $Z_b = 0$ in Eq. (8.14), and the entire effect of the back layers will be due to the impedance of the electrode z_{b1} of thickness t_{b1} so the normalized impedance

$$z_b = jz_{b1} \tan(t_{b1}\gamma) \equiv j \tan \delta \tag{8.19}$$

and the matrix of Eq. (8.14) becomes more complex. The effect of the bond layer and front electrode is even more complex, but an interesting illustrative example of varying the bond layer thickness is shown in Fig. 8.5. For this example, the normalized impedance of the bond layer is taken to be rather low, $z = 0.1$, and it can be seen that even for fairly small thickness the effect on the transducer loss is quite marked. Such a low value of impedance would correspond to the nonmetallic bond materials, but for the metallic bond materials, the impedance mismatch would not be as severe, and the curve of transducer loss would be correspondingly less influenced. This influence of intermediate layers on the shape of the transducer loss curve can be used to determine the band-pass characteristics of the transducer structure. Such impedance transformers can be used, for example, to make the response symmetric about the band center f_0 by making the intermediate layer thickness one-quarter wavelength at f_0. By choosing

Table 8.1 Acoustic Properties of Bond Layer Materials

	Longitudinal waves			Shear waves		
Material	Velocity (cm/sec)	Impedance (g/sec cm^2)	Attenuation (dB/μm @ 1 GHz)	Velocity (cm/sec)	Impedance (g/sec cm^2)	(dB/μm @ 1 GHz)
Epoxy	2.6×10^5	2.86×10^5	Very large	1.22×10^5	1.34×10^5	Very large
Indium	2.25×10^5	16.4×10^5	8	0.91×10^5	6.4×10^5	16
Gold	3.24×10^5	62.5×10^5	0.02	1.2×10^5	23.2	0.1
Silver	5.65×10^5	38×10^5	0.025	1.61×10^5	16.7×10^5	
Aluminum	6.42×10^5	17.3×10^5	0.02	3.04×10^5	8.2×10^5	
Copper	5.01×10^5	40.6×10^5		2.11×10^5	18.3×10^5	

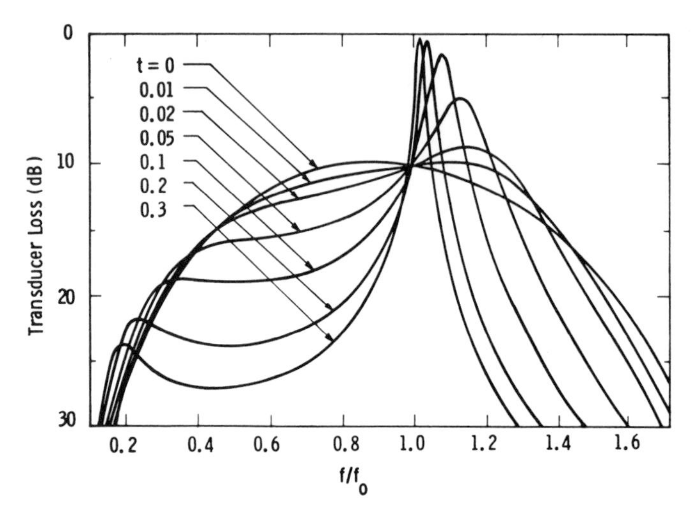

Figure 8.5 Transducer loss for various values of normalized transducer thickness t and intermediate layer normalized thickness 0.1. $R_S = (\omega_0 C_0)^{-1}$, $z_{0t} = 1$, and $\kappa = 0.2$.

other values for the thickness, the bandwidth can be enlarged, ripples smoothed, or various distortions introduced. In general, however, any such objectives are achieved at the expense of increased transducer loss.

8.3 TRANSDUCER MATERIALS

The piezoelectric material itself is perhaps the single most important factor governing the efficiency with which electrical energy can be converted to acoustic energy, this through the electromechanical coupling factor κ. The coupling efficiency is equal to κ^2. Prior to the discovery of lithium niobate, quartz was the most commonly used transducer material, although its coupling factor, even for the most efficient crystal orientations, is rather small. The very high-efficiency transducers were introduced with the discovery of various new ferroelectrics, such as lithium niobate, lithium tantalate, and the ceramic PZT materials, lead-titanate-zirconate. While the PZT transducers have among the highest values of κ, up to 0.7, they are not suitable for high-frequency applications since they cannot be polished to very thin plates. The most suitable piezoelectric transducer materials for high-frequency applications and their important properties are listed in Table 8.2, which is based on a compilation of Meitzler [8].

To produce transducers in the high-frequency range, say larger than 100 MHz, the piezoelectric crystal must be very thin (<20 to 30 μm).

Table 8.2 Properties of Transducer Materials

Material	Density	Mode	Orientation	K	ε_{rel}	V(cm/sec)	Z (g/sec cm^2)
$LiNbO_3$	4.64	L	36° Y	0.49	38.6	7.4×10^5	34.3×10^5
		S	163° Y	0.62	42.9	4.56×10^5	21.2×10^5
		S	X	0.68	44.3	4.8×10^5	22.3×10^5
$LiTaO_3$	7.45	L	47° Y	0.29	42.7	7.4×10^5	55.2×10^5
		S	X	0.44	42.6	4.2×10^5	31.4×10^5
$LiIO_3$	4.5	L	Z	0.51	6	2.5×10^5	11.3×10^5
		S	Y	0.6	8	2.5×10^5	11.3×10^5
$Ba_2NaNb_5O_{15}$	5.41	L	Z	0.57	32	6.2×10^5	33.3×10^5
		S	Y	0.25	227	3.7×10^5	19.8×10^5
$LiGaO_2$	4.19	L	Z	0.30	8.5	6.3×10^5	26.2×10^5
$LiGeO_3$	3.50	L	Z	0.31	12.1	6.5×10^5	22.8×10^5
α-SiO_2	2.65	L	X	0.098	4.58	5.7×10^5	15.2×10^5
		S	Y	0.137	4.58	3.8×10^5	10.2×10^5
ZnO	5.68	L	Z	0.27	8.8	6.4×10^5	36.2×10^5
		S	30° Y	0.35	8.6	3.2×10^5	18.4×10^5
		S	Y	0.31	8.3	2.9×10^5	16.4×10^5
CdS	4.82	L	Z	0.15	9.5	4.5×10^5	21.7×10^5
		S	40° Y	0.21	9.3	2.1×10^5	10.1×10^5
$Bi_{12}GeO_{20}$	9.22	L	[111]	0.19	38.6	3.3×10^5	30.4×10^5
		S	[110]	0.32	38.6	1.8×10^5	16.2×10^5
AlN	3.26	L	Z	0.20	8.5	10.4×10^5	34.0×10^5

There are three well-established techniques for fabricating such thin transducers. In the first method, the piezoelectric plate is lapped to the desired thickness by the usual optic shop methods and then bonded to the delay medium. This method becomes impossibly difficult for transducers of even small area as their frequency increases, because such thin plates cannot be manipulated. A much more convenient technique is to bond the piezoelectric plates with a convenient thickness, say several tenths of a millimeter, to the delay medium and then lap the plate to the final thickness. In both of these methods, one electrode is first deposited on the delay medium, and in the case of thinning the piezoelectric after bonding, the second electrode and back layers are deposited as the final step. Care is required in lapping the bonded transducer so that the base electrode is not damaged by the polishing compound. If a chemically active compound, such as Cyton is used, the delay medium as well as the electrode may be attacked and must be protected by some appropriate coating, such as wax. The final electrical connection to the top electrode must be made in some fashion which does not mass-load the transducer and distort its band-pass characteristics or be so small as to cause hot spots from high current densities. A convenient method is simply to use a small dab of silver print paste to make contact to a wire.

The most successful method for fabricating high-frequency transducers is by deposition of thin films of piezoelectric materials by methods which yield a desired crystallographic orientation for generating either longitudinal or shear acoustic modes [9,10]. The materials used in this method are CdS and ZnO, whose properties are shown in Table 8.2. Such piezoelectric thin films generally cannot be grown with values of κ as high as that of the bulk material, but in the best circumstances κ may approach 90%. Thin-film transducers with bond center frequencies up to 5 GHz can be prepared by these techniques.

A problem that arises with large-area transducers, or even with small-area transducers at very high frequencies, is that of matching the electrical impedance to that of the source. It is especially true for the ferrolectric, piezoelectric materials of very high dielectric constant that the impedance of the transducer may be so low that it becomes difficult to efficiently couple electrical power from the source to the transducer. This problem can be largely overcome by dividing the transducer into a series-connected mosaic, as reported by Weinert and deKlerk [11]. A schematic representation of such a mosaic transducer is shown in Fig. 8.6. If a transducer of given area is divided into N elements which are connected in series, the capacity of the transducer will be reduced by a factor of N^2. An an example, a 1-GHz lithium niobate transducer of 0.25-cm^2 area would represent a capacitive impedance of only 0.038 Ω; if this area were divided into a 16-element mosaic, the impedance would be increased to 10 Ω. A photograph of a 40-element thin-film transducer is shown in Fig. 8.7. The same considerations dictate a mosaic design at low frequencies for

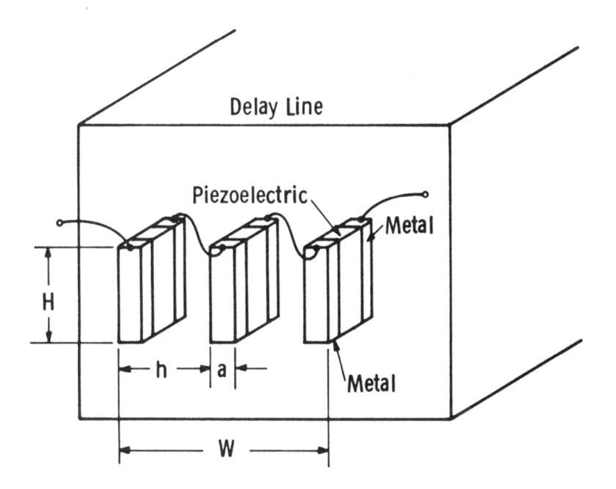

Figure 8.6 Schematic of mosaic transducer.

Figure 8.7 Forty-element thin-film mosaic transducer array. (Courtesy of Westinghouse R&D Center, Pittsburgh, Pa.)

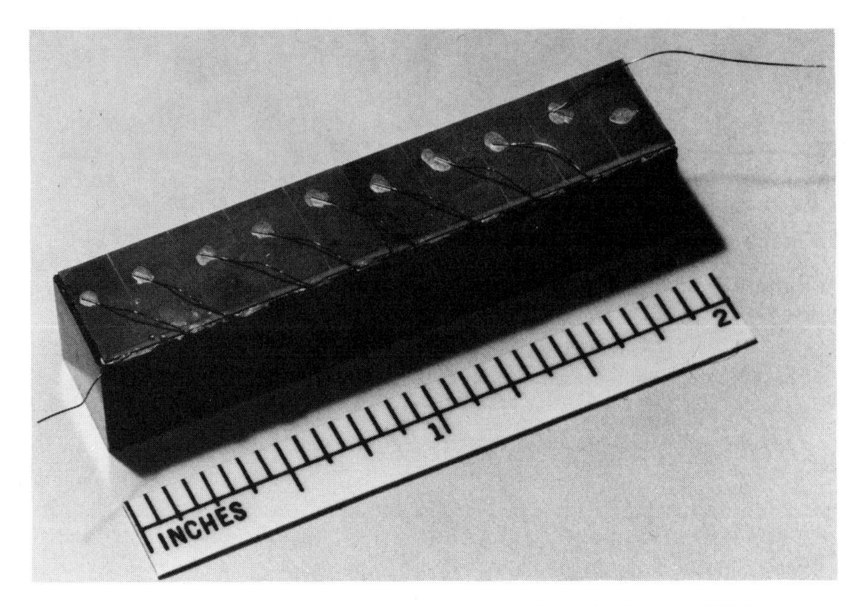

Figure 8.8 Germanium acousto-optic cell with 10-element PZT transducer array. (Courtesy of Westinghouse R&D Center, Pittsburgh, Pa.)

large-area transducers of such materials as PZT, with relative dielectric constants of several hundred. A photograph of a germanium acousto-optic scanner with a 5-cm^2 PZT 13-MHz transducer with 10 elements, series-wired to give 50-Ω impedance, is shown in the photograph in Fig. 8.8.

8.4 ACOUSTIC BEAM STEERING

One of the serious limitations of normal Bragg acousto-optic deflectors is that imposed by the bandwidth, as limited by the Bragg interaction. The most straightforward method of enlarging the interaction bandwidth is simply to shorten the interaction length in order to increase the acoustic beam diffraction spread. This is generally not a very desirable method to increase bandwidth because it is wasteful of acoustic power; only those momentum components of the acoustic beam which can be phase-matched to incident and diffracted light momentum components are useful. Furthermore, as the interaction length shrinks, the transducer becomes increasingly narrow, with a corresponding increase in power density. This increase in power density may produce heating at the transducer, which can cause thermal distortion in the

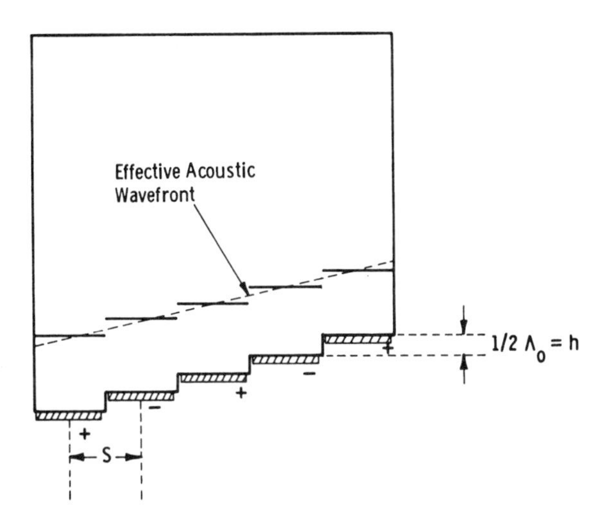

Figure 8.9 Stepped transducer array.

deflector due to gradients in the acoustic velocity and refractive index. An ideal solution to this difficulty would be one in which the acoustic beam changes in direction as the frequency is changed, so that for every frequency the Bragg angle is perfectly matched. The first approximation to such acoustic beam steering was carried out by Korpel [12] for a television display system. This transducer consisted of a stepped array, as shown in Fig. 8.9. The height of each step is one-half an acoustic wavelength at the band center $(1/2)\Lambda_0$, and the spacing between elements s is chosen so as to optimize the tracking of the Bragg angle. Each element is driven π rad out of phase with respect to the adjacent elements, and the net effect of such a transducer is to generate an acoustic wave with corrugated wavefronts, which are tilted at an angle with respect to the transducer surfaces when the frequency differs from the band center frequency f_0. For this transducer configuration, the acoustic beam steers with frequency but matches the Bragg angle only imperfectly. To understand the steering properties of such an acoustic array which was analyzed in detail by Coquin [13], consider the somewhat simpler arrangement shown in Fig. 8.10 in which each transducer element is driven ψ rad out of phase with respect to the next one, and ψ may be electrically varied. This causes the effective wavefront to be tilted by an angle θ_e with respect to the piecewise wavefronts radiating from the individual elements. If θ_e is small, it can be approximated by

$$\theta_e \approx \tan \theta_e = \frac{\psi}{2\pi} \frac{\Lambda}{s} = \frac{\psi}{Ks} \tag{8.20}$$

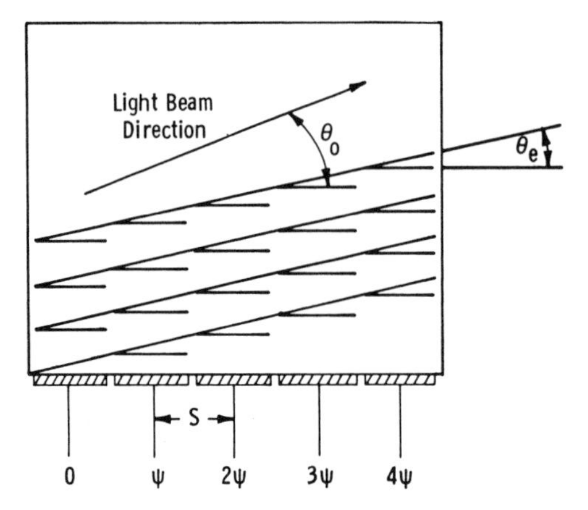

Figure 8.10 Steering of an acoustic beam by a phased array transducer.

If the incident light beam makes an angle θ_0 with the plane of the transducer and if the Bragg angle is $\theta_B = K/2k$, then the angular error from perfect matching is

$$\Delta\theta = (\theta_0 - \theta_e) - \theta_B = \left(\theta_0 - \frac{\psi}{Ks}\right) - \frac{K}{2k} \tag{8.21}$$

The condition for perfect beam steering is that $\Delta\theta = 0$ for all values of K; by setting $\Delta\theta = 0$, the required phase for perfect beam steering is

$$\psi_P = \theta_0 Ks - \frac{K^2}{2k} s \tag{8.22}$$

from which it can be seen that the phase must be a quadratic function of the acoustic frequency. Most of the work done on acoustic beam steering has involved making various approximations to this condition. One such approximation is obtained by making ψ a linear function of frequency, with $\psi = 0$ at f_0, the midband frequency. This was accomplished in the step transducer method of Fig. 8.9 as described in Ref. 12. For this case, the angle that the effective wavefronts make with respect to the transducer plane is

$$\theta_e \cong \frac{\pi}{Ks} - \frac{h}{s} \tag{8.23}$$

where h is the step height and there is 180° phase shift between adjacent elements. The resulting beam-steering error is

$$\Delta\theta = \left(\theta_0 - \frac{K}{2k}\right) + \left(\frac{h}{s} - \frac{\pi}{Ks}\right) \tag{8.24}$$

which can be made zero at the midband frequency f_0 by choosing

$$h = \frac{1}{2}\Lambda_0$$

$$s = \frac{\Lambda_0^2}{\lambda} \tag{8.25}$$

and

$$\theta_0 = \frac{1}{2}\frac{\lambda}{\Lambda_0}$$

A further improvement can be achieved by noting from Eq. (8.23) that θ_e varies as $1/f$, whereas perfect beam steering should lead to a linear variation of θ_e with f. Therefore, the constants h, s, and θ_e may be chosen to agree with the perfect beam-steering case at two frequencies rather than only one, as shown in Fig. 8.11. This first-order beam steering can yield substantial improvements in performance for systems requiring less than a one-octave bandwidth [14], but bandwidths larger than this require a better approximation to the quadratic dependence of the phase on the acoustic frequency.

The next higher approximation to perfect beam steering was carried out by Coquin [13] for a 10-element array, as shown in Fig. 8.12. If the phase applied to each transducer corresponds to that for perfect steering $\psi_\ell = \ell\psi_p$ and the element spacing is $s = \Lambda_0^2/\lambda$, the bandwidth extends from 0 to about $1.6f_0$, the high-frequency drop-off being determined by the finite element spacing. Coquin pointed out that the deflector performance is very tolerant of errors in the individual phases; for example, if the phase applied to each transducer is within 45° of the perfect beam-steering phase, there is a loss of only 0.8 dB in diffracted light intensity. If the phase error is increased to 90°, the loss increases to 3 dB. Thus, for deflectors in which this degree of ripple is permissible, the transducer array may be driven by logic circuitry which sets digital phase shifters. This requires prior knowledge of the input frequency, or analog phase shifters, which accomplish the same function without the need for logic circuits.

An entirely different approach to broad-band Bragg acousto-optic interaction matching is the use of the tilted transducer array, first reported by Eschler [15]. A tilted transducer array consists of two or more transducers electrically connected in parallel and tilted with respect to each other, as illustrated in Fig. 8.13. Each transducer element in the array is designed to cover some fraction of the entire

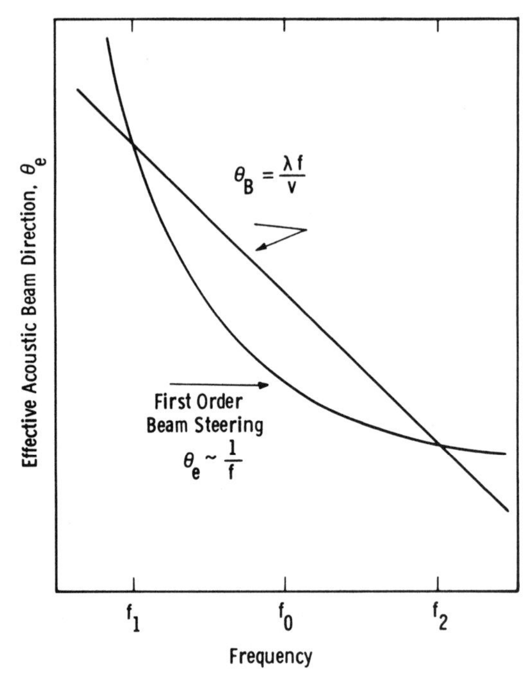

Figure 8.11 First-order beam steering with exact match at two frequencies.

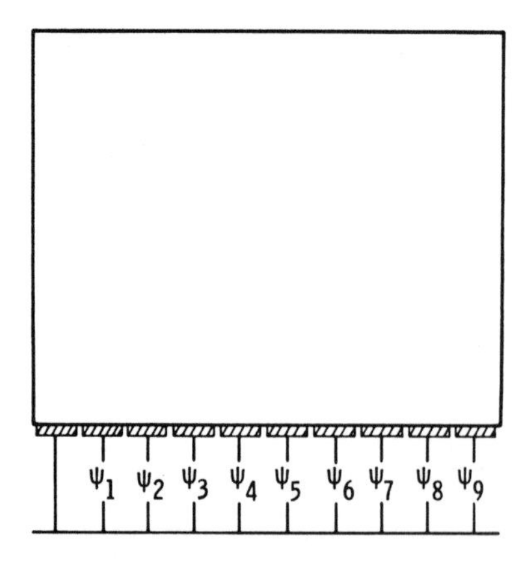

Figure 8.12 Ten-element phased array transducer in which ψ_n = 0°, 180°, or 270°, leading to diffracted intensity less than 0.8 dB lower than for perfect beam steering.

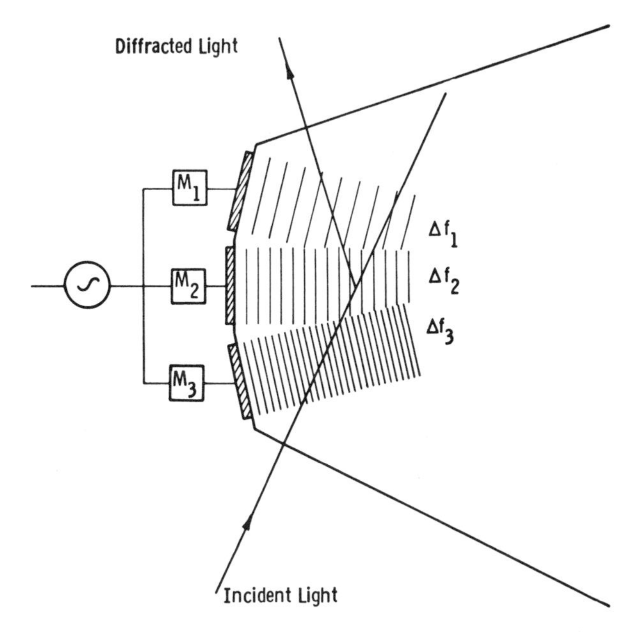

Figure 8.13 Tilted transducer array in which each element is optimized for part of the entire frequency band.

bandwidth, and its angle with respect to the incident light direction is chosen to match the Bragg angle at the center of its subband. For frequencies near the midband of any of the transducer elements, the incident light will interact strongly only with the sound wave emanating from that element; interaction will be weak from the other elements both because the angle of incidence will be mismatched and the frequency will be far from the resonance frequency of those elements. On the other hand, for frequencies which are midway between the resonance frequencies of adjacent elements, that is, $(1/2)(f_{01} + f_{02})$ or $(1/2)(f_{02} + f_{03})$, the contributions to the acoustic fields from both elements are about equal, and the effective wavefront direction lies midway between that of the components. Thus, the array behaves very much as if the acoustic wave were steering with frequency, although this is not true in a strict sense. The diffraction efficiency of the tilted transducer array is shown in Fig. 8.14 in which the solid curves represent the efficiency of the individual elements and the dotted curve represents the overall efficiency. There will typically be about 1 dB of ripple across the full band, which is acceptable for most applications. There are two additional advantages to the tilted array transducer. First, it is obviously relatively simpler to design a larger overall acoustic bandwidth, since each elements of the tilted array need be

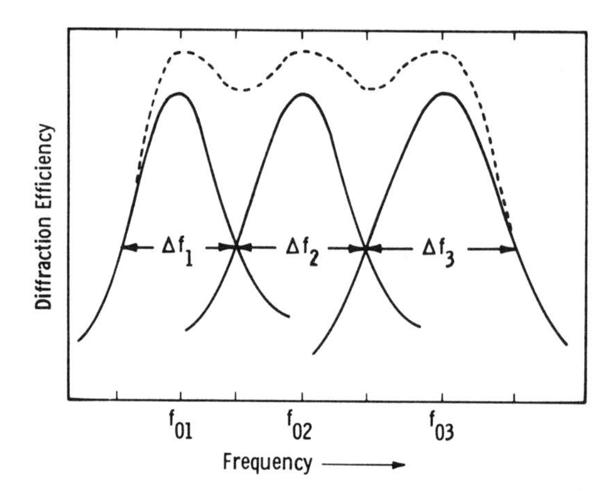

Figure 8.14 Diffraction efficiency of a three-element tilted transducer array. Solid curves represent the efficiency of individual elements, and the dotted curve represents the efficiency of the entire array.

only about one-third of the total bandwidth. Second, tilted array transducers can generally be operated more deeply in the Bragg mode as the combined acoustic wavefronts from adjacent elements are twice the length of that from a single element. If the second-order diffracted light is sufficiently low, then operation over a frequency range larger than one octave is possible. The elements of the array can be connected in parallel since they will tend to behave as band-pass filters, the power being directed to the elements with the closest frequency range. In practice, it is generally necessary to provide impedance-matching networks, as indicated in Fig. 8.13, because of the low reactance obtained with a parallel network.

REFERENCES

1. E. K. Sittig, *IEEE Trans. Sonics Ultrason. SU-16*:2 (1969).
2. A. H. Meitzler and E. K. Sittig, *J. Appl. Phys. 40*:4341 (1969).
3. W. P. Mason, *Electro-mechanical Transducers and Wave Filters*, Van Nostrand-Reinhold, New York (1948).
4. E. K. Sittig and H. D. Cook, *Proc. IEEE 56*:1375 (Aug. 1968).
5. E. K. Sittig, *IEEE Trans. Sonics Ultrason. 16*:2 (1969).
6. W. F. Konog, L. B. Lambert, and D. L. Schilling, *IRE Int. Conv. Rec. Pt. 6, 9*:285 (March 1961).
7. J. D. Larson and D. K. Winslow, *IEEE Trans. Sonics Ultrason. SU-18*:142 (July, 1971).
8. A. H. Meitzler, *Ultrasonic Transducer Materials* (O. E. Mattiat, ed.), Plenum, New York (1971).

9. J. deKlerk, Fabrication of Vapor Deposited Thin Film Piezoelectric Transducers, *Physical Acoustics*, Vol. IV, (W. P. Mason, ed.), Academic, New York (1970), Chap. 5.
10. J. deKlerk, *IEEE Trans. Sonics Ultrason. SU-13*:100 (1966).
11. R. W. Weinert and J. deKlerk, *IEEE Trans. Sonics Ultrason. SU-19*:354 (July 1972).
12. A. Korpel, R. Adler, P. Desmares, and W. Watson, *Proc. IEEE 54*:1429 (Oct. 1966).
13. G. Coquin, J. Griffin, and L. Anderson, *IEEE Trans. Sonics Ultrason. SU-17*:34 (1971).
14. D. A. Pinnow, *IEEE Trans. Sonics Ultrason. SU-18*:209 (Oct. 1971).
15. H. Eschler, *Opt. Commun. 6*:230 (Nov. 1972).

9

Scanning Systems

9.1 APPLICATIONS AND REQUIREMENTS

There are many applications of lasers requiring scanning devices with a very wide range of performance specifications regarding resolution and speed and with the need for sequential scanning or random access addressing. Acousto-optic systems are best suited for those applications in which the cost is moderate, sequential scanning is needed, and the required linear resolution is less than 10^3 spots. Random access addressing can be achieved in times of approximately 10 μsec. Where very fast access times are required and high cost systems are involved, electro-optic scanning methods are better suited. A number of representative applications for which acousto-optic scanning is suitable are listed in Table 9.1. Each application will have its own set of optic requirements with respect to beam shaping and imaging, vertical deflection, intensity modulation, and blanking. The acousto-optically scanned television display was probably the first such application and incorporates most of the required elements of the other systems. Although acousto-optic television displays were described as early as 1939, the first practical system was described by Korpel et al. [1] in 1966. A more recent television rate scanner, incorporating higher-performance materials, was demonstrated by Gorog et al. [2,3] in which the vertical deflection is done by a galvanometer or a rotating prism. The scan of a standard television operates at a rate of 525 lines per frame, with 30 frames per second, which corresponds to a scan time for one line of 6.35×10^{-5} sec and 1.58×10^4 lines per second. These requirements are easily within the capability of acousto-optic scanners.

Before describing the optic system of an acousto-optic scanner, it is important to first mention an important result of linearly swept frequency diffraction. This is the focusing properties of such diffraction, which is easily understood by reference to Fig. 9.1. The focal point of the scanner is located at a distance F given by

$$F \simeq \frac{D}{\theta_1 - \theta_2} \tag{9.1}$$

where D is the aperture and θ_1 and θ_2 are the diffraction angles corresponding to the high and low frequencies f_1 and f_2 within the aperture:

Table 9.1 Acousto-optic Scanning Systems

Display
Large-screen TV
Heads-up cockpit display
Optical radar
Signal processing
Random access optic memories
Laser-addressed transparencies
Information readout
Heat printers
Computer printout
Special tubes
Laser-addressed image tubes (amplifier-intensifier)
Scanning microscope
Laser-scanned orthicon tubes
Manufacturing
Mask fabrication
Pattern cutting
Trimming
Inspection

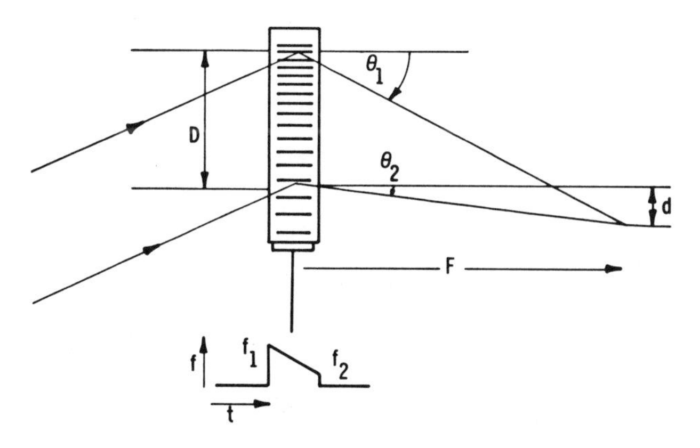

Figure 9.1 Convergence of light diffracted by linearly swept acoustic frequency.

$$\theta_1 = \frac{\lambda f_1}{v}$$
$$\theta_2 = \frac{\lambda f_2}{v} \tag{9.2}$$

If the acoustic travel time across the aperture is $\Delta t = D/v$, then it follows easily that

$$F = \frac{v^2}{\lambda(df/dt)} \tag{9.3}$$

where df/dt is the frequency scan rate. This result was first obtained by Gerig and Montague [4], who also pointed out that the time duration required for the focal spot of width a to traverse a narrow slit in the focal plane is a/v. This phenomenon has been utilized in optical processing of radar signals for performing pulse compression [5,6]. Obviously, since df/dt can be either positive or negative, the diffracted light can be either convergent or divergent. The most straightforward method of compensating for the cylindrical lensing due to this effect is the placement in the optic path of a cylindrical lens with the negative of the focal length calculated from Eq. (9.3).

9.2 ACOUSTO-OPTIC TELEVISION DISPLAY SYSTEM

The scanning system described in Ref. 3 for TV display exemplifies many of the considerations for various types of scanning systems and will be described in some detail here. The optic system, incorporates a galvanometer-type vertical deflector, is shown in Fig. 9.2, and the system incorporating the rotating prism utilizes essentially the same optic train. The components s_1, c_1, and s_2 are used to reshape the small, circular cross-section input laser beam into a large elliptic cross-section beam which provides a good match to the rectangular aperture of the acousto-optic cell. A long, narrow rectangle is required to provide both high efficiency and many resolution elements. Thus, this combination of lenses is a beam expander in which the cylindrical lens c_1 focuses the light to accommodate the narrow dimension of the acousto-optic cell. The combination of lenses s_3 and s_4 is a telephoto arrangement whose long effective focal length produces a magnified image of the scanner at the viewing screen. The cylindrical lens c_2 is used to provide the proper convergence of the light beam to fill the vertical scan mirror aperture, and the last cylindrical lens c_3 is used to project the final image onto the screen. A zero-order stop following the postscan optics prevents unwanted, unmodulated light from reaching the viewing screen. The cylindrical lensing effect

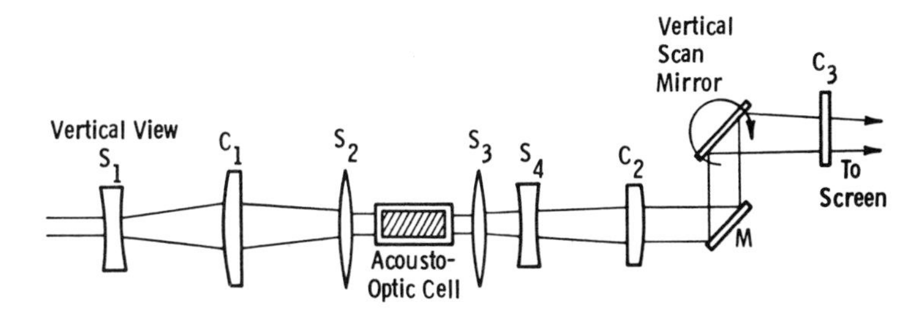

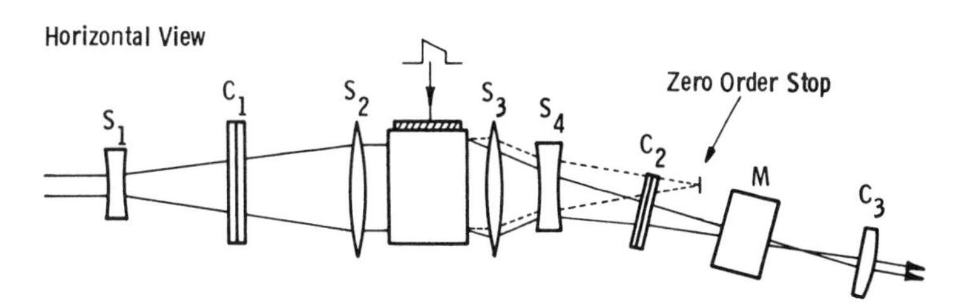

Figure 9.2 Vertical and horizontal views of scanning system for television display. S refers to spherical lenses, and C refers to cylindrical lenses.

described by Eq. 9.3 is compensated by adjustment of the positions of the cylindrical lenses backward or forward along the axis.

The electronic system used by Gorog to drive the acousto-optic scanner was designed for a standard black and white television receiver, as shown in the block diagram of Fig. 9.3. The video chain

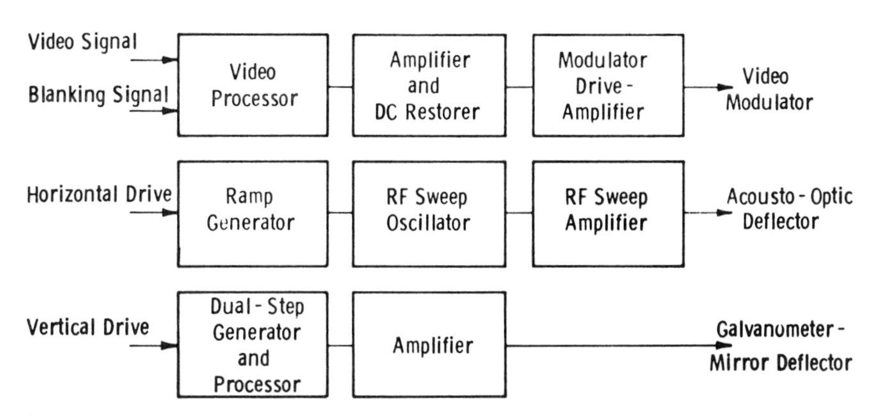

Figure 9.3 Electronic system for television display.

consists of the processor, into which the video and blanking signals are fed, which extracts pure video signal, followed by amplification and dc restoration. The horizontal drive may consist of any number of commercially available swept frequency oscillators followed by a wideband power amplifier, capable of delivering output in the range from 1 to 10 watts, depending on the requirements of the acousto-optic deflector material. The details of the vertical drive will depend on the type of scanner used. For a galvanometer type of scanner, the high mechanical Q would normally result in overshoot and ringing at the end of the trace. The dual-step generator and processor provides current at two intervals and of such amplitude that the galvanometer comes to rest at the end of each sweep.

An important characteristic to consider in acousto-optic scanned displays is the ratio of the flyback time τ to the active line scan time T, since this will affect the resolution of the system. To avoid stray light or other disturbances at the end of each line, the laser beam is blanked at the start of the next line scan until the acoustic wave fills the optic aperture. The minimum flyback time for an acousto-optic scanner is the time required for the acoustic wave to cross the optic aperture, typically on the order of 10 μsec, much larger than the electronic flyback time. Because the laser is blanked for a time duration τ after initiation of the frequency sweep, a fraction of the total frequency sweep is lost, equal to $\tau/(T + \tau)$. Then it is easy to show that the number of resolution elements is reduced from $\tau\ \Delta f$ to

$$N = \left(1 - \frac{\tau}{T + \tau}\right)(\tau\ \Delta f) \tag{9.4}$$

As an example, if the transit time is one-tenth of the scan time, the resulting resolution is 91% of that computed from the time-bandwidth product.

9.3 FLYING SPOT MICROSCOPE

Another type of system to which acousto-optic scanners have been applied with good success is the flying spot laser microscope. Such systems have been reported by Lekavich et al. [7] using visible light and Sherman and Black [8] at infrared wavelengths. The basic function of this device is to scan the focus of the laser beam over the field to be examined, detect the reflected or transmitted light, and display the magnified image on a television receiver. The advantage of a scanned laser flying spot microscope over other types of TV microscopes, such as a vidicon microscope, is that the high-intensity laser beam allows the imaging to be done with potentially a much greater signal to noise ratio. In the latter system, the entire field is illuminated, and the image is projected by conventional optics onto the face

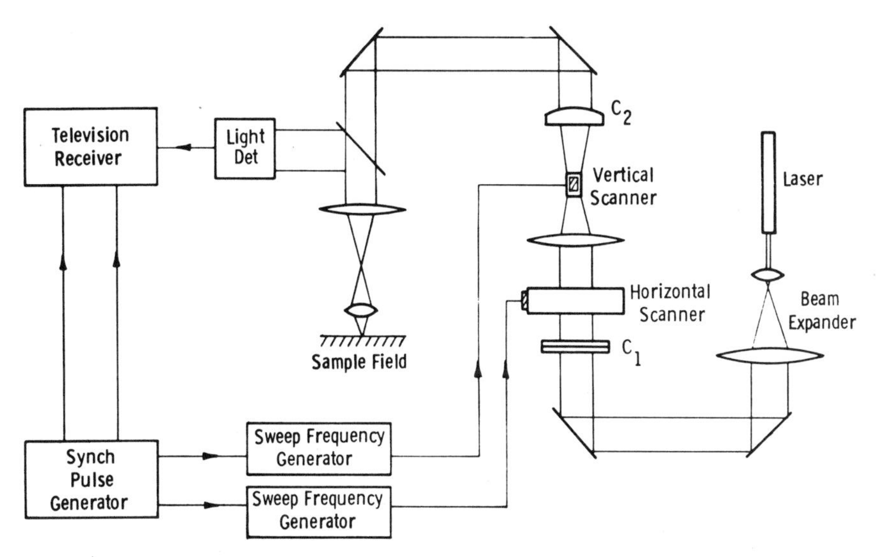

Figure 9.4 Flying spot laser microscope system.

of a vidicon. A scanned electron beam converts this image into an electrical signal for transmission to the television receiver. A block diagram of a scanned laser flying spot microscope, as reported in Ref. 7, is shown in Fig. 9.4. The beam from a 1-mW helium-neon laser is expanded by the telescope to 1.5-cm diameter, and the cylindrical lens c_1 is used to shape the beam to fill the long, narrow aperture of the first acousto-optic scanner. This scanner produces the horizontal deflection, and the spherical lens following the first scanner rotates the beam to match the aperture of the vertical scanner. The cylindrical lens c_2 restores the beam to its original circular cross-section, after which the microscope optics bring it to a focus in the plane of the sample field, which is scanned by the focused spot. The horizontal Bragg scanner operates at the line rate of the television receiver, 15.75 kHz, and the vertical Bragg scanner operates at the standard frame rate of 1/30 sec. The light reflected by the sample field falls on the photodetector where it is converted into an appropriate electrical signal for the television receiver. The synch pulse generator provides the timing pulses for the television receiver and also for the sweep frequency oscillators which drive the horizontal and vertical Bragg scanners. In the work reported in Ref. 7, the horizontal scan frequency ran from 18 to 38 MHz in 53.5 μsec with 10-μsec blanking time. The vertical scan frequency ran from 20.5 to 35.5 MHz in 15.3 msec with 1.3-msec blanking time. Because relatively low bandwidths and frequencies were required, water was used as the acousto-optic medium.

The arrangement for viewing the object, as shown in Fig. 9.4, uses a beam splitter for on-axis reflection. Other viewing methods include off-axis reflection, in which no beam splitter is required, resulting in lower light losses. Improved depth perception is reported but with lower signal to noise ratio, because of the small-angle light scattering. The magnification of the microscope is the ratio of the width of the TV screen to the width of the scanned field:

$$M = \frac{W}{\alpha F} \tag{9.5}$$

where $\alpha = \lambda N/D$ is the maximum scan angle and F is the focal length of the final lens. If the f number of the lens is f, then

$$M = \frac{W}{f\lambda N} \tag{9.6}$$

To illustrate the limitations on the magnification, W = 40 cm, the maximum aperture diffraction-limited lens might correspond to an f number of 1.5, and the number of resolution elements for good image quality is no less than 200, so that at $\lambda = 0.6328$ μm, the resulting magnification is M = 2100. In practice, much larger f numbers are used, with correspondingly lower magnification.

The x-y deflection system described here utilizes two separate Bragg cells conveniently displaced along the optic axis so that the oblong beam cross-section could be rotated through 90°. There may, however, be applications where the required resolution and deflection efficiency can be obtained with a circular optic beam profile. For such a case, there is no reason for any separation between the x and y deflectors, and it may be more economical to combine both x and y deflectors in a single piece of material rather than two separate ones. It had been believed that the simultaneous x and y deflection of a beam of light was necessarily less efficient than cascaded deflectors in which no acoustic fields could overlap. An analysis by LaMachia and Coquin [9] showed that both simultaneous and cascaded deflection are capable of identical efficiency and that the intensity of diffracted light in overlapping acoustic beams is exactly the same as from two physically separated acoustic beams. A problem may result from different acoustic velocities and diffraction efficiencies in orthogonal directions in the same crystal, but generally these will not be so drastically different as to make operation impossible. The choice of cascaded or simultaneous deflection can then be made entirely on the basis of effective design and economy.

9.4 ACOUSTO-OPTIC ARRAY SCANNERS

There are many applications of scanned laser systems where the requirements on the number of resolution elements or bandwidth are greater than the state of the art capability of single acousto-optic devices. In such cases it may be possible to satisfy the system requirements by channelizing the device into a multiplicity of scanners. There are obviously many ways in which this may be done, depending on the functions of the system. Two examples will be described here, one in which the bandwidth is multiplied and the other in which the resolution is multiplied by array configurations.

A linear acousto-optic array was reported by Young and Yao [10] for application as a page composer for wideband recording. This device consists of a large number of individually driven transducers mounted on an acousto-optic medium, as illustrated in Fig. 9.5. Such units were built with up to 138 channels, with total recorder bandwidths up to 750 Mbits/sec. For a large number of channels to be placed on a crystal of convenient size, the transducers must be of small width and located very close to each other. This will result in the acoustic fields of neighboring channels overlapping each other by acoustic diffraction a short distance from the transducer. To avoid such cross-talk, the incident light must be focused onto the modulator array within a distance d from the transducers, given approximately by

$$d \approx \frac{1}{2} \frac{sHf}{v} \tag{9.7}$$

where H is the width of the transducers and s is the spacing between them. This requirement can be rather severe, as shown by the following example. Suppose the modulator crystal is 25 mm long, and we

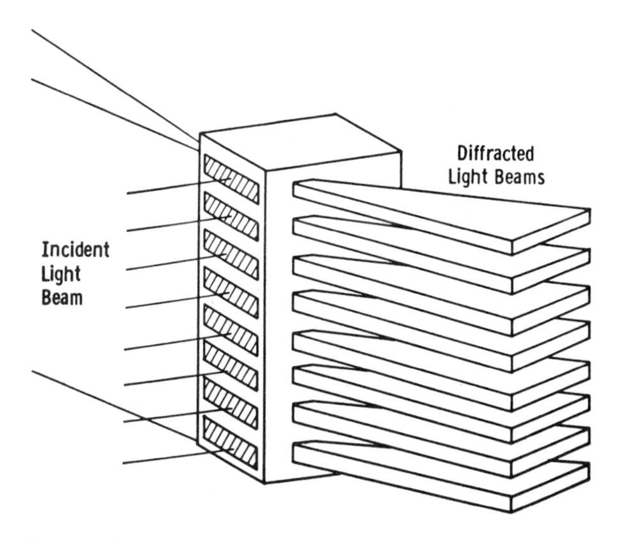

Figure 9.5 Linear acousto-optic scanner array for page composer.

wish to place 100 channels on the crystal; if the center frequency is 100 MHz, then for an acoustic velocity of 3×10^5 cm/sec, d is 2.6×10^{-2} cm. Careful design is required to achieve a high optic throughput; in particular it is important to match the acoustic and optic beam spreads. Results were reported in Ref. 10 for devices fabricated from SF-8, SF-59, and TeO_2 glasses and $PbMoO_4$ and TeO_2 crystals. Electrical power as low as 26 mW per channel were obtained using the fast shear mode in TeO_2.

In another type of acousto-optic array, the objective is to multiply the number of resolution elements available from a single channel, so that the elements of the array are arranged in series, rather than in parallel, as was the case with the previous example. Many recording applications require resolution of 1500 points per line or better, and it is not now possible to achieve this performance with a single scanner. An optic arrangement has been proposed [11] in which an array of scanners, each with a resolution capability of 500 points per line, together yield 1500 points per line. The optic system for producing a linear scan from an array is rather complex, with the input light beam following a helical path so as to pass through the acoustic field of each of the transducers on successive passes through the device. The deflected beam from each scanner is extracted and aligned by a set of prisms so as to give a single continuous scan on the recording medium. Very high resolutions are, in principle, possible if the optic complexity can be dealt with.

9.5 ACOUSTO-OPTIC REFRACTIVE EFFECTS FOR RESOLUTION ENHANCEMENT

The great majority of acousto-optic devices and applications use the diffraction effects that have been discussed in this chapter. There is, however, another class of acousto-optic effects based on the acoustically induced refraction of light. While the same physical mechanisms are involved in both effects, the ratio of light beam width to acoustic wavelength determines whether diffraction or refraction predominates. When this ratio $w/\Lambda \ll 1$, refraction will take place, while for $w/\Lambda \gg 1$, diffraction takes place. Several of the refraction effects that are easily observed at low acoustic frequencies are illustrated in Fig. 9.6. Part (a) of the figure represents the cylindrical lensing effect due to the refractive index gradient set up by the acoustic field. Such lens-like behavior was proposed by DeMaria and Danielson [12] as a method for internal laser modulation, and the calculated trajectories of light rays crossing through an acoustic field were reported. From these calculations it was determined that paraxial rays are brought to a fairly good focus provided that the optic aperture is no greater than approximately $0.3\ \Lambda$.

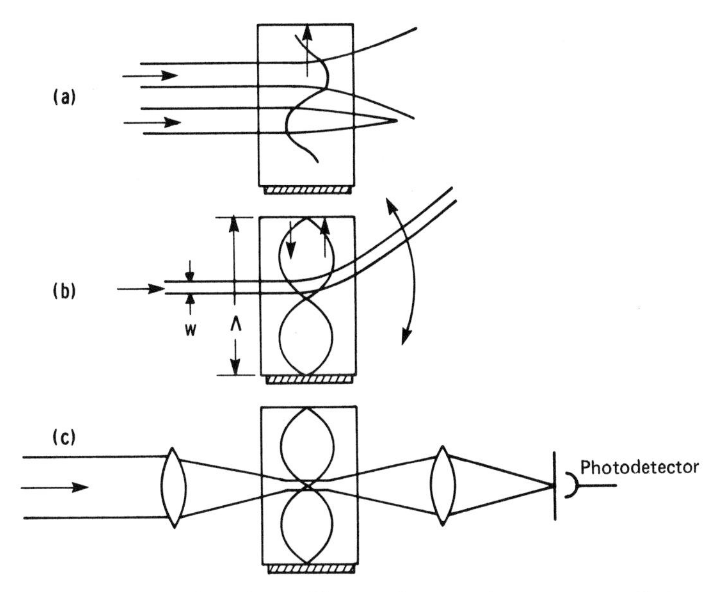

Figure 9.6 Refraction effects of acoustic waves on light when $w/\Lambda << 1$. (a) Positive and negative cylindrical lensing, (b) light beam scanning, (c) amplitude modulation of light beam.

Part (b) of Fig. 9.6 shows the effects of ray bending when a light beam of width $w << \Lambda$ passes through the sound field on one side of a node of a standing acoustic wave. This approximates a linear refractive index gradient, and the light beam is deflected through an angle which varies sinusoidally with time and whose peak is determined by the acoustic standing-wave intensity. The use of ray optics easily leads to an evaluation of the maximum deflection angle, as measured in air external to the modulator, with the result

$$\theta_{max} = 2\pi \frac{L}{\Lambda} \Delta n \tag{9.8}$$

where Δn is the peak refractive index change. To obtain large deflection angles, very large acoustic power densities are required. These can be achieved by operation in high Q resonant acoustic cavities. Such deflection was demonstrated by Lipnick et al. [13] and Aas and Erf [14], who achieved peak-to-peak deflections as large as 6° at a frequency of 320 kHz.

Part (c) of Fig. 9.6 illustrates a method in which refraction can be used not only to simply amplitude-modulate a light beam but also to image the intensity distribution of a standing-wave acoustic field [15]. The light beam is brought to a focus within the acoustic field by the first lens and then refocused by a second lens onto a spatially filtered photodetector. When the acoustic intensity is zero, all the light passes through the aperture onto the photodetector. When the acoustic intensity rises, the cylindrical lensing effect tends to defocus the light reaching the spatial filter. Since this occurs twice for each period of the acoustic standing wave, the modulated light frequency will be double that of the acoustic wave. A useful property of all the refractive phenomena described above are that they are inherently achromatic, the wavelength dependence being entirely due to the dispersion of the medium in which they take place. Thus, acousto-optic refractive devices are very well suited for use with broad-band light sources.

A novel method for enhancing the resolution achievable with a Bragg-type deflection cell utilizes a traveling-wave acoustic lens. The underlying ideas were first proposed by Foster et al. [16] and later extended by Yao et al. [17] to a guided acoustic wave version of the traveling-wave lens in order to reduce the power requirements. The basic arrangement for using an acoustic traveling-wave lens to increase the resolution of an acousto-optic scanner is illustrated in Fig. 9.7. The underlying idea is to provide the system with a final converging lens which remains at all times axial with the beam as it is being scanned. The increase in resolution that such a system provides is the ratio of the diameter of the input beam to the diameter of the output beam. A moving lens must be used rather than a stationary lens, since it is necessary to maintain axial alignment. Otherwise, the focused spot position will remain fixed as the beam is scanned. By using the refractive effects described above, a synchronously moving lens can be realized by passing the scanning beam through a pulse burst of appropriately chosen center frequency.

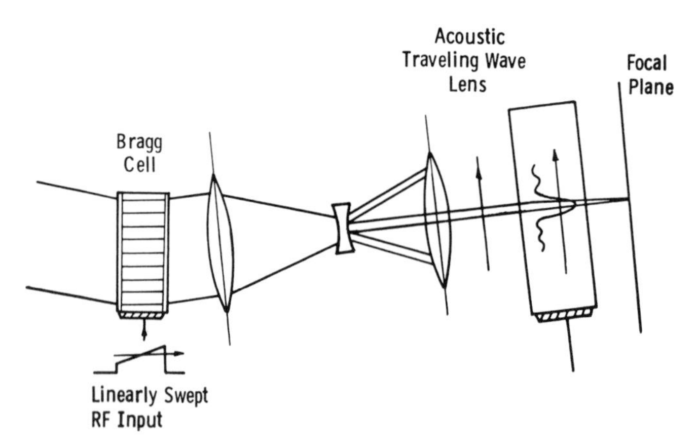

Figure 9.7 Enhancement of scanner resolution with acoustic traveling-wave lens.

A simple analysis based on the ray trace approach of DeMaria and Davidson [12] was used by Foster et al. [16] to derive the resolution enhancement factor of an acoustic traveling-wave lens. The analysis assumes a collimated input beam with a Gaussian intensity distribution of wavelength λ and a beam waist diameter w_0. If this beam is focused by an acoustic lens of focal length F, the beam waist diameter of the focused spot is [18]

$$w_1 \approx \frac{\lambda F}{\pi w_0} \tag{9.9}$$

If the input beam and lens are traveling, then for each resolution element of the input beam of diameter w_0 there will be w_0/w_1 resolution elements of the output beam. Thus, the ratio of input to output resolution element number is

$$\frac{N_1}{N_0} = \frac{w_0}{w_1} = \frac{\pi w_0^2}{\lambda F} \tag{9.10}$$

The ray trace analysis indicates that the largest aperture for which reasonably good focusing properties hold is $\Lambda/4$, so that the input beam diameter and acoustic frequency are chosen to satisfy

$$w_0 = \frac{\Lambda}{4} = \frac{v}{4f} \tag{9.11}$$

For the paraxial focal length the ray trace analysis of the acoustic lens yields the expression

$$F = \frac{1}{4} \Lambda \sqrt{\frac{n_0}{\Delta n}} \tag{9.12}$$

so that substitution of this equation and Eq. (9.11) into Eq. (9.10) yields the result

$$\frac{N_1}{N_0} = \frac{\pi \Lambda}{4\lambda} \left(\frac{\Delta n}{n_0}\right)^{1/2} \tag{9.13}$$

for the resolution enhancement factor. This may be related to the acoustic power density through Eq. (6.21), from which we obtain

$$P_A = \frac{2}{M_2} (\Delta n)^2 \tag{9.14}$$

and

$$\frac{N_1}{N_0} = \frac{\pi\Lambda}{4\lambda}\left(\frac{P_A M_2}{2n_0^2}\right)^{1/4} \tag{9.15}$$

As an example of the performance to be expected from such a system, assume a scanner with an input resolution of $N_0 = 200$ having an input beam diameter of 1 mm. If we assume an acousto-optic material with an acoustic velocity of 3×10^5 cm/sec, the frequency for which $w_0 = \Lambda/4$ is 750 kHz. The largest refractive index changes that are generally possible to reach are on the order of $\Delta n = 10^{-4}$, which for a refractive index $n_0 = 2$ corresponds to an acoustic lens focal length of F = 14.4 cm. For a wavelength $\lambda = 0.633$ μm, Eq. (9.10) predicts a resolution enhancement factor of 35, so that the final resolution of the system would be $N_1 = 7000$, far in excess of that possible with any Bragg cell scanner alone. For a material with an acoustic-optic figure of merit 100 times larger than that of fused quartz, $M_2 = 1.6 \times 10^{-16}$ $\sec^3/g$, and an acoustic power density of 12.5 W/cm^2 will be required to produce $\Delta n = 10^{-4}$. This would be possible only on a pulse basis.

9.6 MODULATION TRANSFER CHARACTERISTICS OF ACOUSTO-OPTIC SCANNERS

An important consideration for high-resolution deflection systems is the frequency dependence of the response. This is determined by the modulator aperture and intensity distribution of the light beam as well as other factors. The optical transfer function (OTF) of an incoherent, diffraction-limited opic system was derived by Goodman [19] and applied to the acousto-optic deflector by Randolph and Morrison [20]. The frequency response of an acousto-optic scanner under several simplifying assumptions is easily derived. Consider a one-dimensional scan in the x direction and a diffraction-limited light beam of aperture width D with an intensity distribution of the light in the focal plane, which can be considered as being of infinitesimal size, provided that it is much smaller than the smallest spatial frequency in the modulator. If the modulation function applied to the light beam is M(t), the instantaneous intensity in the focal plane is

$$I(x,t) = M(t)S[x - x'(t)] \tag{9.16}$$

where $x'(t)$ is the instantaneous position of the beam. The static intensity distribution of the light beam is denoted by $S[x - x'(t)]$. The integrated energy falling on the focal plane is

$$E(x) = \int_{-\infty}^{+\infty} I(x,t)\,dt$$

measurable either with detectors or visually as with a CRT display.

For a constant modulation frequency ω and an acousto-optic medium with acoustic velocity v,

$$M(t) = e^{i\omega t} \tag{9.18}$$

and

$$x'(t) = vt \tag{9.19}$$

For a uniformly illuminated aperture of width D,

$$S(x - x') = \frac{\sin(\pi D/\lambda F)(x - vt)}{(\pi D/\lambda F)(x - vt)} \tag{9.20}$$

where F is the distance to the focal plane. With the substitution

$$\zeta = x - vt \tag{9.21}$$

we obtain

$$E(x) = \frac{1}{v} e^{i(\omega x/v)} \int_{-\infty}^{+\infty} e^{-i(\omega\zeta/v)} \left| \frac{\sin(\pi D/\lambda F)\zeta}{(\pi D/\lambda F)\zeta} \right|^2 d\zeta \tag{9.22}$$

The normalized response of the scanner is

$$R(\omega) = \frac{E(\omega)}{E(0)} = 1 - \frac{\omega}{\omega_{max}} \tag{9.23}$$

where

$$\omega_{max} = 2\pi \frac{Dv}{\lambda F} \tag{9.24}$$

obtained by carrying out the integration in Eq. (9.22).

In the more general case, the illumination of the scanner aperture is not uniform but is a truncated Gaussian as would be the case for the lowest-order-mode laser beam. The modulation transfer function can be evaluated following the formalism of Refs. 19 and 20 for an aperture illumination function given by

$$P(z) = e^{-(z/z_0)^2} \tag{9.25}$$

where z is the coordinate along the modulator aperture and z_0 is the point of $1/e^2$ intensity. The modulation transfer function has been shown to be equivalent to the autocorrelation function of P(z), with the result

$$R(\omega) = \frac{\int_{-(1/2)[D-(\lambda F\omega/2\pi v)]}^{(1/2)[D-(\lambda F\omega/2\pi v)]} \exp\{-2[z^2 + (\lambda F\omega/4\pi v)^2]/z_0^2\}\, dz}{\int_{-(1/2)D}^{(1/2)D} \exp[-(z/z_0)^2]\, dz} \tag{9.26}$$

The geometric interpretation of this integral has been given as illustrated in Fig. 9.8, which represents the formation of the autocorrelation function of the aperture light intensity distribution with a relative shift of $\lambda F\omega/2\pi v$. This description greatly simplifies the evaluation of the integral for a truncated Gaussian light beam. It is easily seen from Fig. 9.8 that the integral is nonzero only when the shifted distributions have a region of overlap. This requires that

$$\frac{\lambda F\omega}{4\pi v} < \frac{D}{2} \tag{9.27}$$

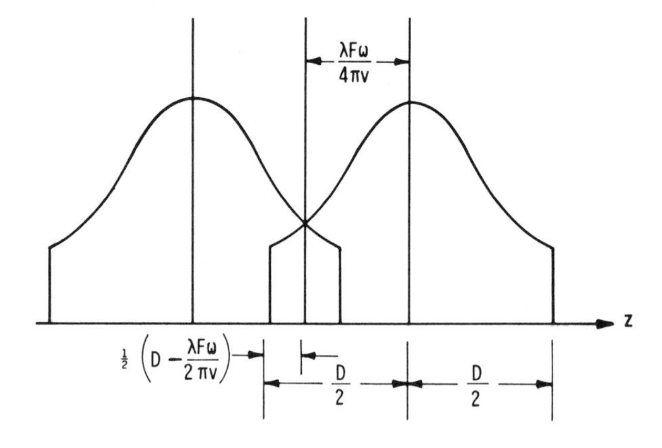

Figure 9.8 Geometric interpretation of the modulation transfer function as the autocorrelation function of the aperture light distribution shifted by $(1/2)[D - (\lambda F\omega/2\pi v)]$.

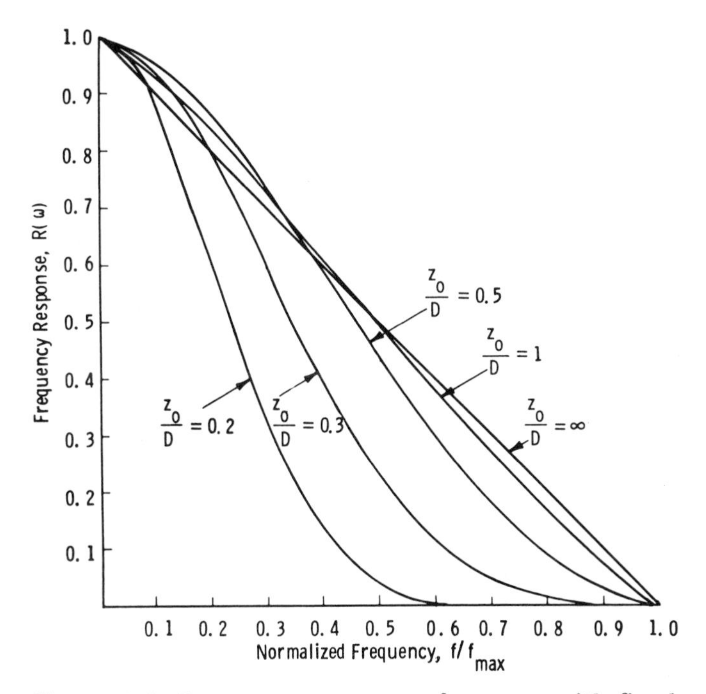

Figure 9.9 Frequency response of scanner with fixed aperture D illuminated by Gaussian beam of $1/e^2$ half-width z_0.

or that the maximum frequency for nonzero response is

$$f_{max} = \frac{\omega_{max}}{2\pi} = \frac{Dv}{\lambda F} \tag{9.28}$$

The response functions are easily computed with the aid of a programmable calculator and provide some insight into the effects of relative size of aperture and Gaussian beam width. Plots of the frequency response of a scanner with fixed aperture D and various values of z_0 are shown in Fig. 9.9. The curve for $z_0/D = \infty$ (uniform illumination) corresponds to Eq. (9.23), and it can be seen that as z_0 is decreased, the high-frequency response becomes increasingly degraded. At the same time, there is an improvement in the low-frequency response. The effect of changing aperture size for a constant beam width z_0 is shown in Fig. 9.10 in which the normal frequency is taken relative to the maximum frequency for an aperture $D = z_0$. It is clear from these curves that when the resolution is limited by the scanner aperture it is desirable to have a large light beam width but that there is not much improvement in the response of the system once the width is equal in

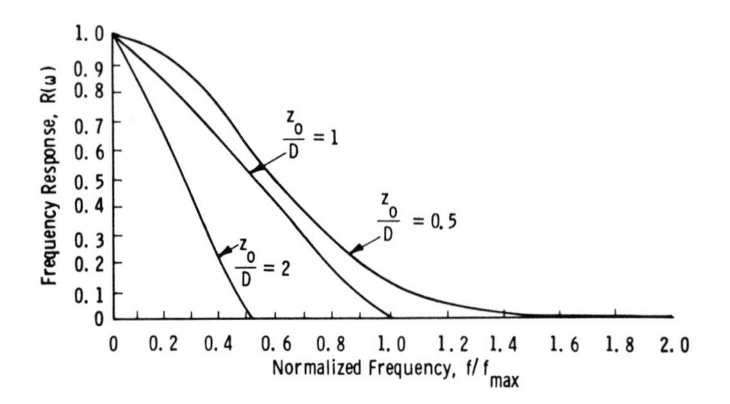

Figure 9.10 Frequency response of scanner illuminated with Gaussian beam of fixed width z_0. f_{max} corresponds to aperture $D = z_0$.

size to the aperture. On the other hand, if the resolution is limited by the beam width, the response will always be improved by increasing the aperture. However, there will not be much increase in high-frequency response for apertures larger than twice the beam width.

REFERENCES

1. A. Korpel, R. Adler, P. Desmares, and W. Watson, *Appl. Opt.* *5*:1667 (Oct. 1966).
2. I. Gorog, J. D. Knox, and P. V. Goedertier, *RCA Rev.* *33*:623 (Dec. 1972).
3. I. Gorog, J. D. Knox, P. V. Geordertier, and I. Shidlovsky, *RCA Rev.* *33*:667 (Dec. 1972).
4. J. S. Gerig and H. Montague, *Proc. IEEE* *52*:1753 (1964).
5. M. B. Schulz, M. G. Holland, and L. Davis, *Appl. Phys. Lett.* *11*:237 (1967).
6. J. H. Collins, E. G. Lean, and H. J. Shaw, *Appl. Phys. Lett.* *11*:240 (1967).
7. J. Lekavich, G. Hrbek, and W. Watson, Flying Spot Laser Microscope, *Proc. Electro-Optical Syst. Design Conf., NY (Sept. 1970)*, Industrial and Scientific Conference Management, Inc., Chicago, p. 650.
8. B. Sherman and J. F. Black, *Appl. Opt.* *9*:802 (April 1970).
9. J. T. LaMachia and G. A. Coquin, *Proc. IEEE* *36*:304 (Feb. 1971).
10. E. H. Young and S. K. Yao, *Linear Array Acoustooptic Devices*, *Proc. IEEE Ultrason. Symp., Annapolis* (Sept. 1976), p. 666.
11. J. T. McNaney, *Laser Focus*, 84 (June 1979).

12. A. J. DeMaria and G. E. Danielson, *IEEE J. Quantum Electron. QE2*:157 (July 1966).
13. R. Lipnick, A. Reich, and G. Schoen, *Proc. IEEE (Corres.) 52*:853 (July 1964).
14. H. Aas and R. K. Erf, *J. Acoust. Soc. Am. 36*:1906 (Oct. 1964).
15. A. Korpel and L. Kessler, Acoustic Holography by Optically Sampling a Sound Field in Bulb, *Acoustic Holography*, Vol. 2 (A. F. Metherell and L. Larmore, eds.), Plenum, New York (1970), Chap. 9.
16. L. C. Foster, C. B. Crubly, and R. L. Cohoon, *Appl. Opt. 9*:2154 (Sept. 1970).
17. S. K. Yao, D. Weid, and R. M. Montgomery, *Appl. Opt. 18*:446 (Feb. 15, 1979).
18. R. L. Fork, D. R. Herriott, and A. Kogelnik, *Appl. Opt. 3*:1471 (1964).
19. J. W. Goodman, *Introduction to Fourier Optics*, McGraw-Hill, New York (1968), p. 116.
20. J. Randolph and J. Morrison, *Appl. Opt. 10*:1383 (June 1971).

10

Acousto-Optic Light Diffraction in Thin Films

10.1 OPTIC WAVEGUIDES

A recent trend in optics has been the development of thin-film optic devices in which light beams are confined to propagate in layers of higher refractive index than the substrate on which they are deposited. This has led to the relatively new field of integrated optics in which complex optic systems may be incorporated onto a small *optic chip*. The advantages of such an approach include a high degree of miniaturization and the possibility of inexpensive batch fabrication, leading to the same economies that integrated electronic circuits conferred upon the electronics field. So far, these goals are nowhere near being realized, but active research is in progress to solve the difficult fabrication problems. It is not an objective of this book to present a detailed account of optic waveguide devices; there are several excellent reviews [1-3] which summarize this field for the reader who wishes more information. We shall merely state here some of the most basic characteristics as they will be required to understand scanning devices. The most common planar waveguide is asymmetric, being bounded on one side generally by air, or a medium of low refractive index n_1, and on the other side by a substrate of refractive index n_3, intermediate between its own index n_2 and that of the superstrate, $n_1 < n_2 < n_3$. Its thickness t is on the order of the wavelength of light, or a small multiple thereof, so that only one or a few discrete modes of propagation may be supported. Figure 10.1 shows the qualitative variations of the field of the light for the fundamental and first guided modes for the planar waveguide. Some fraction of the energy in the mode is carried outside the guide layer itself, in the substrate and superstrate. This fraction will depend on the relative guide thickness t/λ and the relative indices of the three component media. This energy, external to the guide layer, is termed the evanescent wave and is normally small, unless n_2 and n_3 are very close in value or t/λ is very small. There are two basic types of waveguide propagation modes, TE and TM. The difference is essentially the polarization; for all TM modes the polarization (i.e., E field) is normal to the plane of the wavefuige, while for all TE modes it lies in the waveguide plane.

The wavelength of a guided wave and its propagation constant $\beta = 2\pi/\lambda$, are complex functions of the guide thickness and refractive indices of the materials. Figure 10.2 shows a typical variation of the

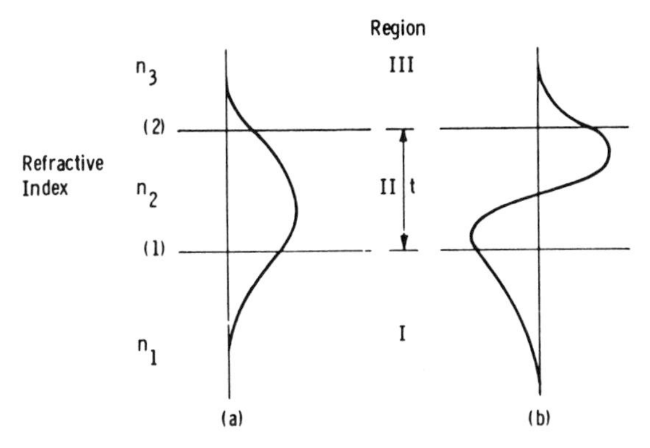

Figure 10.1 Electric field distribution in guided modes in a planar asymmetric waveguide when indices of refraction n_1, n_2, and n_3 are different. For guided wave propagation n_2 must be larger than the other indices; in this asymmetric case, the substrate is considered the material with the larger index of n_1 and n_3. Note that in the evanescent regions I and III the decay of the field is different depending on the index.

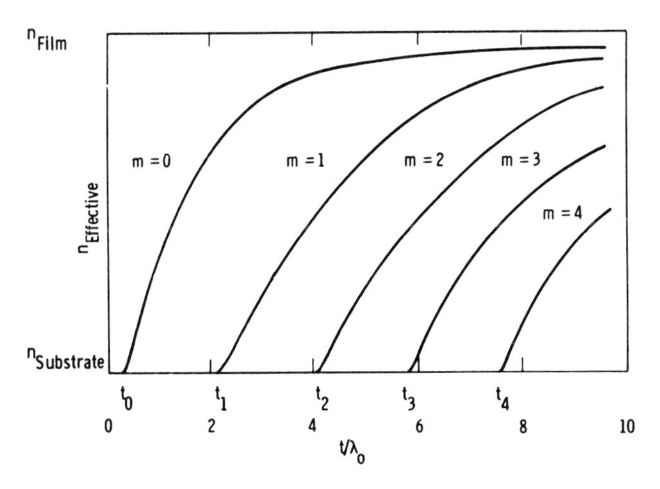

Figure 10.2 Effective index of refraction for asymmetric film waveguides. As the thickness increases, the effective index increases from $n_{substrate}$ to n_{film}.

effective index of refraction of the guided wave n_{eff} as a function of guide thickness, in light wavelengths, for an asymmetric guide. As film thickness increases from the cutoff thickness, the effective index increases from a value asymptotically equal to the bulk index of the guide film material. The effective indices for the modes of only one of the two polarizations, either TE or TM, are shown in Fig. 10.2. The index curves for the TE and TM modes of a given order are slightly displaced from each other, even in isotropic materials, but this small difference is important to the operation of some waveguide devices.

Direct coupling of laser energy into the waveguide by focusing is not efficient because of the small size of the guide. The most useful techniques use coupling from the evanescent field of a prism held in close contact with the guide. Other successful techniques use gratings fabricated on the guide, or tapered structures. More recently, good coupling efficiency has been achieved with a diode laser butt-coupled to the edge of the waveguide. This approach is more consistent than the others, which typically use gas lasers, with the objective of miniaturizing the resulting optic systems. Ultimately, it would be most desirable to incorporate a thin-film laser directly into the waveguide, but there is not yet available a suitable medium, which can be pumped conveniently, that is compatible with commonly used dielectric waveguide systems. These include lithium niobate, in which a surface layer is given a higher refractive index than the underlying material, and high-index glass, which is sputter-deposited onto a lower-index glass, such as quartz. Very high-quality planar waveguides have been made with these materials, and the major objective of integrated optics is the fabrication of several either monolithic or hybrid devices onto a common structure to form an integrated circuit. The devices that we shall be concerned with here are the acousto-optic scanning and deflection structures, which are crucial to a variety of optic signal processing systems.

10.2 ACOUSTIC SURFACE WAVES

The optic energy in a guided mode is confined to a thin surface layer, and so it is readily apparent that efficient interaction with acoustic energy requires that it too be confined to the same surface layer. This occurs naturally with acoustic surface waves (ASW), the technology of which has recently developed very rapidly [4,5]. There are several types of surface waves, defined by the character of the material motion. For the phenomena to be described here, the Rayleigh wave is the most important mode of propagation. The Rayleigh wave is illustrated in Fig. 10.3 with its depth-dependent strains. For isotropic media, the material motion is perpendicular to the propagation

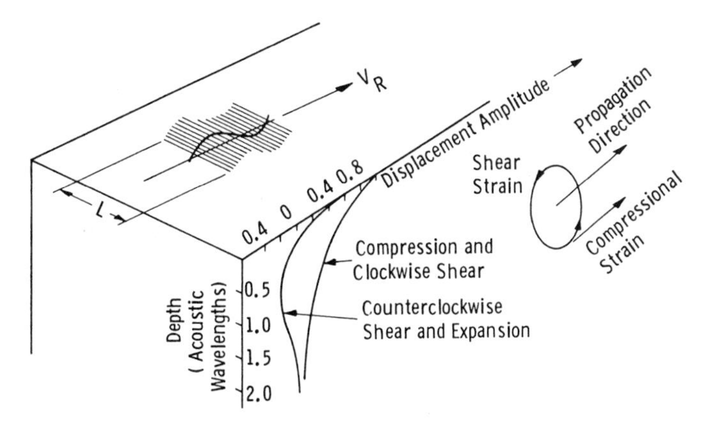

Figure 10.3 Rayleigh wave propagating in a layered medium. Displacements of the shear and compressional components are shown varying with depth below the surface.

surface and may be resolved into a shear displacement normal to the surface and a compressional component along the wave normal. These displacements are

$$u_1 = U_1 \cos(kx_1 - v_R t) \tag{10.1}$$

and

$$u_3 = U_3 \sin(kx_1 - v_R t) \tag{10.2}$$

where v_R is the Rayleigh wave velocity. For an isotropic medium, the Rayleigh wave velocity is given approximately by

$$v_R = \left(\frac{0.87 + 1.12\nu}{1 + \nu} \right) v_s \tag{10.3}$$

where

$$\nu = \frac{v_L^2 - 2v_s^2}{2(v_L^2 - v_s^2)} = \text{Poisson's ratio} \tag{10.4}$$

and v_s and v_L are the bulk material shear and longitudinal wave velocities, respectively. This formula applies well to such optic waveguide materials as sputtered glass but must be used only with great care with crystalline waveguides, such as lithium niobate. The

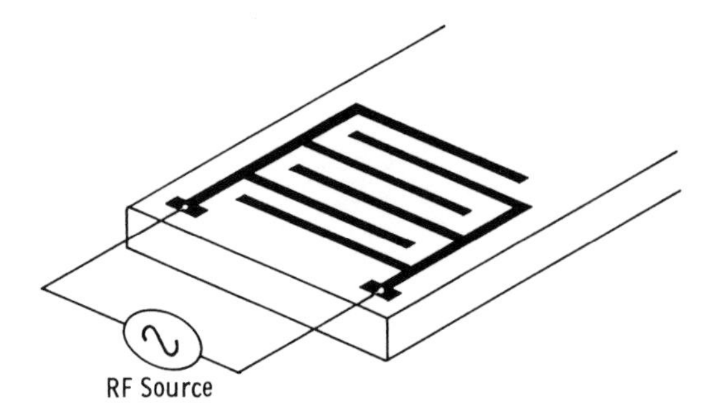

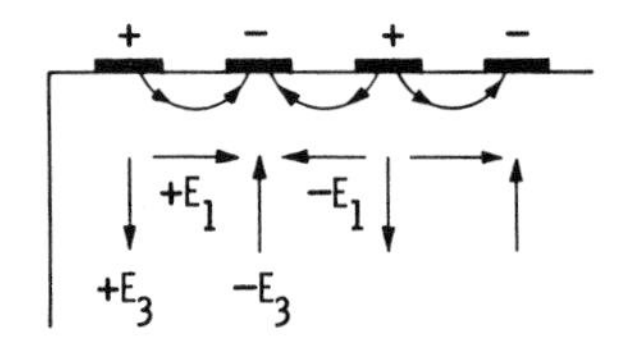

Figure 10.4 (a) Interdigital structure on a piezoelectric substrate for generating surface waves. (b) Electric field distribution within the material produced by the electrode grid.

amplitude of the displacements u decreases exponentially with depth beneath the propagation surface, so that the wave is almost entirely contained within a layer about two acoustic wavelengths thick. The acoustic power flow in the wave is given by

$$P_a = \frac{1}{2} \rho v^2 v_R L e^2 \int_0^H u^2 \, dz \tag{10.5}$$

where H is the depth of the acoustic disturbance, ρ is the density of the material, L is the acoustic beam width, e is the strain amplitude, and v is either the bulk shear or longitudinal wave velocity, appropriate to the shear or longitudinal strain, respectively.

There are a number of ways of generating acoustic surface waves, but the most useful method is by means of an interdigital transducer deposited on a piezoelectric substrate, which might or might not also be the propagation medium. A schematic of an interdigital transducer for launching acoustic surface waves is shown in Fig. 10.4. The electric field distribution which such a grid produces in the piezoelectric material is also shown in the figure. An ac voltage applied between the two comb electrodes produces the field indicated, which has components parallel and normal to the propagation surface; these produce the two component displacements of the Rayleigh wave, which are in phase quadrature. Of course, the orientation of the piezoelectric surface must be chosen so that the two components of the electric field couple to the elastic moduli that will generate the desired shear and longitudinal strains.

10.3 GUIDED WAVE INTERACTIONS WITH ACOUSTIC SURFACE WAVES

Much of the motivation for the development of guided wave-acousto-optic devices has been related to factors such as miniaturization and potential economics of batch fabrication; it is also true, however, that the planar approach may simply lend itself more readily to improved performance of acousto-optic modulators. The basic reason for this lies in the dependence of efficiency on the ratio of interaction length to acoustic beam height, as indicated in Eq. (6.27). Optic waveguide acousto-optic devices can be made much more efficient than bulk devices because diffraction spreading prevents long interaction lengths with narrow beam heights in bulk, while this is not a limitation in planar devices, where the waves are confined in dimension over long distances. Thus, power densities orders of magnitude larger than in bulk can be achieved. It is difficult to fabricate bulk devices with the aspect ratio L/H larger than about 10, so that low power deflectors can only be made from such high-figure-of-merit materials as TeO_2. For waveguide deflectors, L/H may easily range from 100 to 1000, so that lower figure-of-merit materials, such as $LiNbO_3$ or even SiO_2, can be used to make efficient deflectors.

The first demonstration of Bragg diffraction of a guided optic beam by an acoustic surface wave was carried out by Kuhn et al. [6] in a sputtered glass film using an input prism coupler configuration and acoustic waves near 200 MHz. Exactly as for the bulk counterpart, efficient diffraction occurs when the Bragg condition is satisfied. The periodic variation in refractive index is set up by the acoustic surface wave in three possible ways. These are the photoelastic effect, the electro-optic effect, and the corrugations produced at the waveguide interface. The photoelastic contribution dominates in all amorphous materials and most crystals, while the electro-optic effect may dominate in ferroelectric crystals, such as lithium niobate [7].

One of the potentially most useful aspects of acousto-optic diffraction in waveguides relates to anisotropic effects. As shown in Chap. 6, such effects in bulk interactions are determined by the birefringence of the crystal. In optic waveguides, on the other hand, the birefringence is the difference in the effective refractive indices between the two guided light modes that are coupled by the acoustic surface wave. Thus, a guided light beam may be diffracted into the same mode (in which case there is no birefringence), into a mode of the same polarization (TE to TE or TM to TM) but different mode order, or into a mode of different polarization (TE to TM or TM to TE). In the last two cases, the birefringence will be determined not only by the waveguide materials but also by the film thickness. Anisotropic interactions can therefore occur even in isotropic films, over a continuously controlled range of values, governed by the waveguide thickness and the mode orders. As is the case with bulk anisotropic diffraction, very wideband, high-efficiency diffraction may be produced in a waveguide by acoustically coupling two dissimilar optical modes at a center frequency given by

$$f_0 = \frac{v}{\lambda}(N_i^2 - N_d^2)^{1/2} \tag{10.6}$$

where N_i and N_d are the effective refractive indices of the incident and diffracted modes, respectively. This frequency can be chosen to be in a convenient range, say between 10 and 100 MHz, while in bulk crystals, with the exception of TeO_2, the corresponding frequency is in the gigahertz range, as fixed by the crystal birefringence.

A calculation of the efficiency with which acoustic surface waves diffract guided optic waves must take into account the depth dependence of both the acoustic and optic power density. A schematic of the interaction region is shown in Fig. 10.5 in which t is the waveguide thickness and H is the depth below the surface to which the acoustic strain may be considered uniforn. The amplitude is

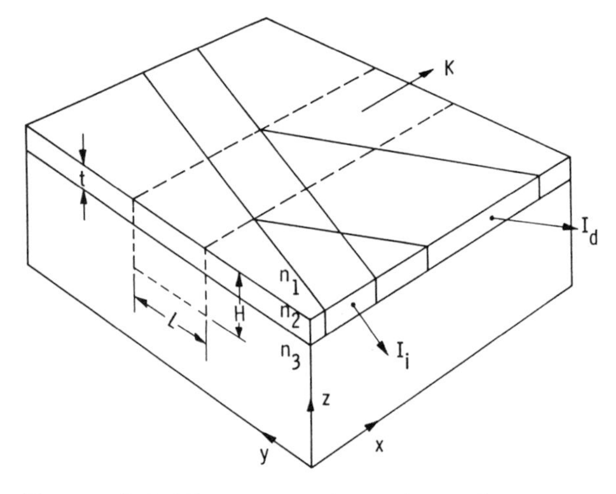

Figure 10.5 Planar acousto-optic surface wave deflector with waveguide thickness t and acoustic wave depth H.

$$e_0 = \left(\frac{2P_a}{LH\rho v^3} \right)^{1/2} \tag{10.7}$$

This strain, in turn, produces a periodic variation in the dielectric constant:

$$\delta \varepsilon = -\varepsilon^2 p e_0 = -Np \left(\frac{2P_a}{HL\rho v^3} \right)^{1/2} \tag{10.8}$$

where, as before, p is the photoelastic constant with suppressed tensor notation. The diffraction efficiency,

$$\frac{I_d}{I_i} = \sin^2 \kappa_{id} L \tag{10.9}$$

where i and d denote incident and diffracted beam parameters, can be determined by evaluating the coupling factor κ_{id} between the incident and diffracted beams. The coupling is given by the integral [1]

$$\kappa_{id} = \frac{\omega \varepsilon_0}{2 \cos \theta_i} \int_{-\infty}^{+\infty} \overline{E}_i^* \overline{\overline{\delta \varepsilon}}\, \overline{E}_d \, dy \tag{10.10}$$

in which $\overline{E}_i$ and $\overline{E}_d$ are the matrices representing the electric fields of the incident and diffracted light, and the variation in the dielectric permittivity tensor is

$$\overline{\overline{\Delta \varepsilon}} = \overline{\overline{\delta \varepsilon}} \cos(\Omega t - Kx) \tag{10.11}$$

By simplifying the tensor quantities to scalar form, the coupling coefficient becomes [8]

$$\kappa_{id} = \frac{k}{2 \cos \theta_i} \left(\frac{2P_a}{HL} \frac{N^6 p^2}{\rho v^3} q_{id}^2 \right)^{1/2} \tag{10.12}$$

where

$$q_{id} = \frac{2kN}{\omega \mu_0} \int_0^{\infty} E_i^*(y) E_d(y) \, dy \tag{10.13}$$

contains the overlap integral for the incident and diffracted beams.

An explicit evaluation of the coupling coefficient has been carried out [8-10]. The formulation due to Schmidt provides useful simplifications for many cases of practical interest. The coupling coefficient due to the surface corrugations for TE modes of the same order reduces to

$$\kappa_{id} = \left(\frac{\pi}{\lambda \cos \theta_i}\right)\left(\frac{n_2^2 - N^2}{N t_{eff}}\right)[U_y(0) - U_y(t)] \qquad (10.14)$$

where the *effective waveguide thickness* is

$$t_{eff} = t + \frac{1}{k(N^2 - n_1^2)^{1/2}} + \frac{1}{k(N - n_3^2)^{1/2}}$$

The effective waveguide thickness is larger than its physical thickness due to the evanescent field penetration of the guided light into the substrate and superstrate. This can be appreciable for weakly guided modes (either $t/\lambda < 1$ or $n_2 \approx n_3$). The amplitude of the corrugation at the superstrate and substrate boundaries are $U_y(0)$ and $U_y(t)$, so that the coupling coefficient due to the corrugation will not be large if the displacement of the air (or superstrate)-waveguide interface is nearly the same as that of the waveguide-substrate interface. This will generally be the case for acoustic wavelengths that are long in comparison with the guide thickness.

Evaluation of the coupling factor is, in general, quite complex, and great care must be taken in determining the diffraction efficiency for any set of conditions [11]. In particular, the value of the overlap integral will depend on the polarization and mode order of both incident and diffracted beams, and the depth dependence of the acoustic surface wave strains will cause the perturbation of the dielectric permittivity as well as the photoelastic constants also to vary with depth.

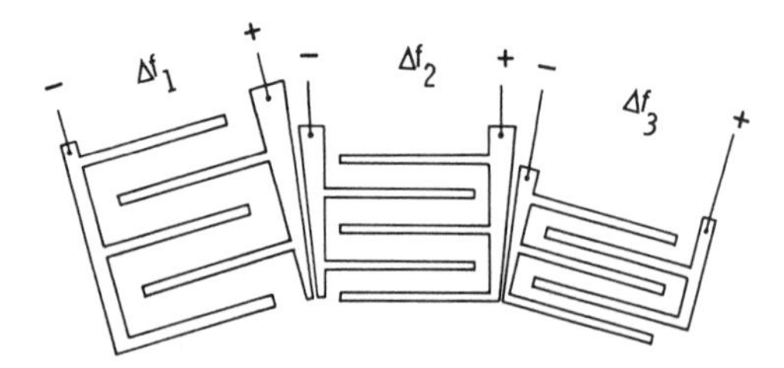

Figure 10.6 Multiple-tilted-element acoustic surface-wave transducer array with staggered center frequencies.

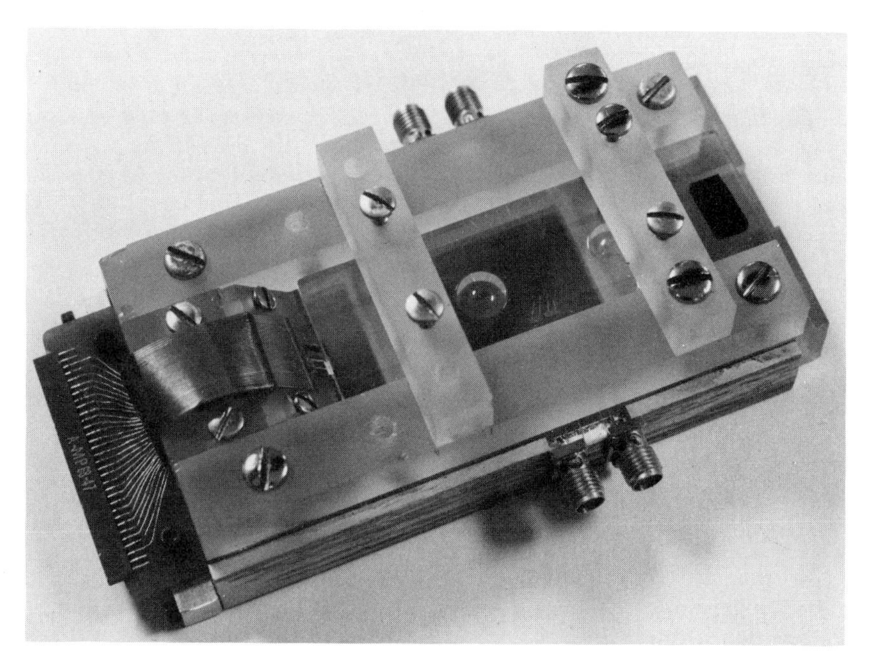

Figure 10.7 Integrated-optic spectrum analyzer. The substrate is lithium niobate with in-diffused waveguide, and geodesic lenses are ground into the substrate. The diode laser and photodetector array are edge-coupled to the waveguide. (Courtesy of Advanced Technology Laboratory, Westinghouse Electric Corp.)

The criteria for selecting materials for planar acousto-optic interactions are similar to those for bulk devices. They must be of exceptionally high optic quality, since the requirements for good transmission through waveguides are extremely severe. Low acoustic losses are needed for wideband operation at high frequencies, and high figures of merit generally require large refractive index and low acoustic velocity. High-efficiency acousto-optic waveguide modulators have been made in deposited films of the chalcogenides As_2S_3 [9] and As_2S_5 [12], Ta_2O_3 [13], and amorphous TeO_2 [14]. Currently, the most useful optic waveguide modulator medium is $LiNbO_3$, in which the high-index waveguide layer is formed by in-diffusion of a metal such as titanium [15].

Integrated-optic devices have proven difficult to bring beyond the proof-of-principle stage because of the problems associated with fabricating all the components on a single chip. Nevertheless, progress has been made in the realization of an integrated-optic spectrum analyzer [16] which utilizes the acousto-optic deflection of a guided light beam into 100 resolution elements by acoustic surface waves in the range from 250 to 500 MHz. Normal Bragg diffraction is utilized; the required interaction bandwidth is achieved with a multiple-element transducer array [17] in which each element, active over only a portion of the full band, is tilted in angle with respect to the incident light beam so as to be optimized over its subband. Such a transducer is illustrated in Fig. 10.6, and the complete spectrum analyzer into which it has been incorporated is shown in Fig. 10.7. A diode laser is end-coupled to the waveguide, and geodesic lenses to collimate and refocus the light have been ground into the lithium niobate. A detector array is butt-coupled to the substrate in the focal plane of the second lens. This spectrum analyzer represents a first effort at reducing to practice a very complex integrated-optic system. As the state of the art progresses and the light source and detector can be more appropriately incorporated onto the substrate, the resulting systems will be truer to the concept of fully integrated-optic circuits.

10.4 BULK ACOUSTIC WAVE INTERACTIONS WITH GUIDED WAVES

Guided optic waves are capable of interacting not only with acoustic surface waves but with bulk waves as well [18]. This is accomplished by either depositing or bonding a transducer on top of the waveguide under which the light beam will pass. A buffer layer, generally a 1-μ-thick film of SiO_2, is first deposited on the waveguide to provide optic isolation from the bottom electrode of the transducer. The acoustic wavelength will be, in general, large compared with the guide thickness, so that instantaneously the refractive index under the transducer will be uniform. Temporally, this refractive index will change according to

$$n = n_0 + \Delta n \sin \omega t \tag{10.15}$$

where ω is the radian frequency of the acoustic wave and Δn is the peak change of refractive index produced by the wave. The principal effect of the acoustic field will be to phase-modulate the light with modulation index

$$\Delta \phi = \frac{2\pi L \, \Delta n \sin \omega t}{\lambda} = - \frac{\pi L n^3 pe \sin \omega t}{\lambda} \tag{10.16}$$

where L is the light propagation path length under the transducer.

It is possible to use the interaction with bulk acoustic waves to make a digital light beam deflector by segmenting the transducer so that the acoustic wave produces a spatially periodic region which acts

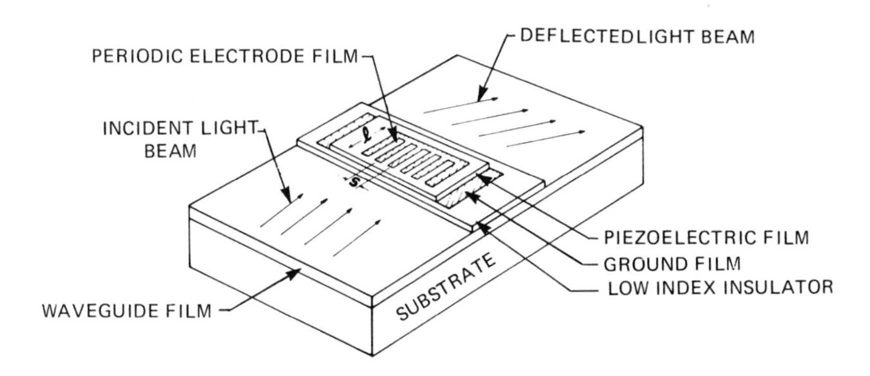

Figure 10.8 Bulk acoustic wave integrated optic sanner.

as a phase grating across the guided light beam wavefront [19]. The geometry of this type of deflector is shown in Fig. 10.8 and is very similar to the electro-optic Bragg waveguide deflector [20]. If the length of the transducer is large in comparison to the spacing s between elements of the transducer, that is, if

$$L >> \frac{s^2}{2\pi \lambda_0} \qquad (10.17)$$

a thick grating is formed, which has the property of diffracting light into a single diffraction order. The intensity in this order will be large provided the Bragg condition is satisfied and that the incident and diffracted angles of the light are given by

$$\sin \psi = \frac{\lambda}{s} \qquad (10.18)$$

There are two important advantages to this type of waveguide modulator. The maximum modulation frequency is determined by acoustic attenuation, or the time required for the acoustic wave to traverse the optic aperture. Since bulk acoustic waves need only pass through the waveguide layer thickness, the attenuation limit is reached at frequencies as high as 50 GHz for high-quality glass waveguides. The other limit to the bandwidth is determined by the transit time of the acoustic wave through the guide layer, so that

$$f_{max} \simeq \frac{v}{t} \qquad (10.19)$$

For single-mode guides, this limit will lie in the 10- to 20-GHz region, although a practical limit will be imposed by the state of the art of high-frequency thin-film transducer fabrication technology. If the bandwidth of this type of modulator is limited by transit time, then it is easy to show that the power per unit bandwidth (for half-wave retardation) is

$$\frac{P}{\Delta f} = \frac{\pi}{4}\left(\frac{w}{L}\right)\left(\frac{\lambda^2}{\kappa^2}\right)\left(\frac{t}{v}\right)\frac{1}{M_2} \tag{10.20}$$

where w is the width of the transducer (designed to match the optic beam aperture) and κ^2 is the electromechanical conversion efficiency of the transducer. The other important advantage of acoustic modulation is that the choice of guide materials is much greater than for electro-optically modulated waveguides. Good acousto-optic materials can be deposited on substrates of choice; waveguide films of As_2S_3 or Tl_3AsS_4, for example, require powers of only several hundred nanowatts per megahertz for 1-cm interaction length.

The structure so far described is capable of deflecting a light beam into just one direction, determined by the transducer element spacing and the light wavelength. It is well known in radar antenna theory that a radiating beam from a phased array may be steered in direction by applying a linearly phased electrical excitation across the elements of the array. In a similar fashion the elements of the transducer depicted in Fig. 10.8 can be connected with a linear phase delay across the array so that the angle through which the light beam is deflected will be determined by the ratio λ/s and also by the phase slope between elements [21]. By adjusting this phase, the light beam can be steered in direction. To achieve a large number of resolution elements, a large phase shift is required, and this has not yet been satisfactorily demonstrated. Some possible methods for producing the phase shift include voltage-controlled Schottky-barrier diodes of variable capacity or techniques for linearly varying the peak voltage at each transducer element. Whatever method is used to produce the phase delay between elements, the light beam diffracted by the array will have the following properties:

1. If there are N elements in the array, the main lobe of the diffracted beam can be directed to N resolvable beam positions.
2. When a linear phase slope of P complete cycles is applied across the array so that the phase of the nth element is

$$\delta_n = 2\pi\left(\frac{nP}{N}\right)$$

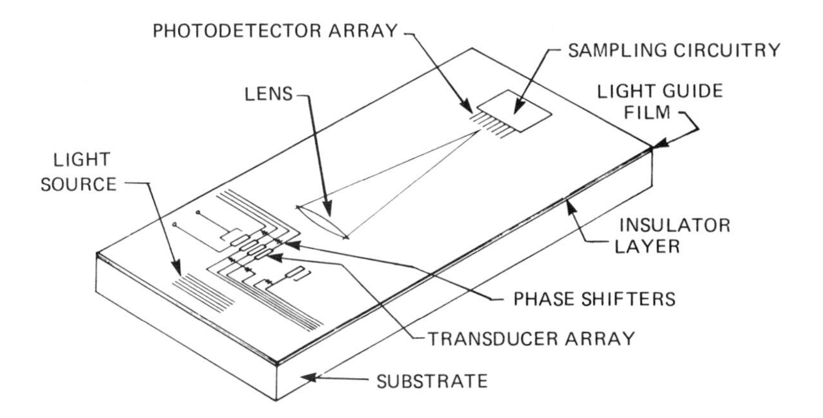

Figure 10.9 Integrated optic A/D converter using bulk acoustic wave scanning.

the angular position of the main lobe of the first diffraction order is deflected to an angle ψ given by

$$\psi \approx \frac{\lambda}{s}\left(1 + \frac{P}{N}\right) \tag{10.21}$$

Since the modulation is sinusoidal with time, having the frequency of the acoustic excitation, ψ will also be sinusoidal, with an amplitude given by the above expression.

The phased array form of the bulk acoustic wave waveguide deflector can be used in systems requiring random access scanning. For example, an integrated optic A/D converter has been proposed [21] in which the electrical phases applied to the transducer array vary in response to the amplitude of the incoming signal. This, in turn, directs the light beam to the elements of a photodetector array whose output represents the digitized signal. The accuracy of this A/D converter is $2^N - 1$ bits, where N is the number of resolvable beam positions; the sampling rate corresponds to acoustic frequency, which can be in the gigahertz range. A schematic representation of the integrated optic A/D converter is shown in Fig. 10.9.

REFERENCES

1. P. K. Tien, *Appl. Opt. 10*:2395 (Nov. 1971).
2. H. Kogelnik, IEEE Trans. *Microwave Theory Tech. MTT-23*:2 (Jan. 1975).
3. P. K. Tien, *Rev. Mod. Phys. 49*:361 (April 1977).
4. A. A. Oliner, ed., *Surface Acoustic Waves*, Springer, Berlin (1979).
5. R. M. White, *Proc. IEEE 58*:1238 (Aug. 1970).
6. L. Kuhn, M. L. Dakss, P. F. Heidrich, and B. A. Scott, *Appl. Phys. Lett 17*:265 (Sept. 1970).
7. J. M. White, P. F. Heidrich, and E. G. Lean, *Electron. Lett. 10*:510 (Nov. 1975).
8. R. V. Schmidt, *IEEE Trans. Sonics Ultrason. SU-23*:22 (1976).
9. Y. Ohmachi, *J. Appl. Phys. 44*:3928 (1973).
10. E. Lean, J. M. White, and C. D. W. Wilkinson, *Proc. IEEE 64*:779 (1976).
11. T. G. Giallorenzi and A. F. Milton, *J. Appl. Phys. 45*:1762 (April 1974).
12. M. Gottlieb and T. J. Isaacs, *Appl. Opt. 17*:2482 (1978).
13. D. A. Wille and M. C. Hamilton, *Appl. Phys. Lett. 24*:159 (Feb. 1974).
14. Y. Ohmachi, *Electron. Lett. 9*:539 (Nov. 15, 1973).
15. R. V. Schmidt and I. P. Kaminow, *IEEE J. Quantum Electron. QE-11*:57 (Jan. 1975).
16. D. Mergerian, E. Malarkey, R. Patenius, J. Bradley, G. Marx, L. Hutcheson, and A. Kellner, *Appl. Opt. 19*:3033 (1980).
17. C. S. Tsai, Le T. Nguyen, S. K. Yao, and M. A. Alhaider, *Appl. Phys. Lett. 26*:140 (Feb. 15, 1975).
18. G. Brandt, M. Gottlieb, and J. Conroy, *Appl. Phys. Lett. 23*:53 (1973).
19. M. Gottlieb and G. Brandt, *Proc. IEEE Ultrason. Symp.*, 231 (1976).
20. J. M. Hammer, D. J. Channin, and M. T. Duffy, *Appl. Phys. Lett. 23*:176 (1973).
21. M. Gottlieb and G. Brandt, *Electron. Lett. 16*:359 (May 1980).

LIST OF SYMBOLS

c	Elastic stiffness coefficient
D	Optic aperture
e	Strain amplitude
f	Frequency of acoustic wave
h	Planck's constant
$\underline{K}$	Momentum vector of acoustic wave
$\underline{k}$	Momentum vector of light wave
L	Acoustic field interaction length
M	Acousto-optic figure of merit
n	Refractive index
P	Acoustic power
p	Photoelastic constant
v	Acoustic wave velocity
w	Beam waist diameter of focused beam
Z	mechanical impedance of acoustic medium
z	Normalized mechanical impedance of acoustic medium
α	Ultrasonic attenuation coefficient, dB/cm
β	Compressibility
Γ	Ultrasonic attenuation constant, dB/cm $(MHz)^2$
γ	Normalized acoustic frequency
ε	Dielectric constant
θ_B	Bragg angle
κ	Electromechanical coupling factor
Λ	Sound wavelength
λ	Light wavelength
ρ	Density
τ	Travel time of acoustic across aperture
Ω	Radian frequency of acoustic wave
ω	Radian frequency of light

Index